Adsorption From Aqueous Solutions

Adsorption From Aqueous Solutions

Edited by

P.H. TEWARI

Atomic Energy of Canada Limited
Pinawa, Manitoba, Canada

PLENUM PRESS • NEW YORK AND LONDON

Library of Congress Cataloging in Publication Data

Main entry under title:

Adsorption from aqueous solutions.
  "Proceedings of a symposium . . . held March 24-27, 1980, as a satellite symposium to the
meeting of the American Chemical Society Division of Colloid and Surface Chemistry,
Houston, Texas."
  Bibliography: p.
  Includes index.
  Contents: The power exchange function / Donald Langmuir — The TiO_2/aqueous elec-
trolyte system / R. O. James, P.J. Stiglich, and T.W. Healy — Effects of strong binding of
anionic adsorbates on adsorption of trace metals on amorphous iron oxyhydroxide / Mark
M. Benjamin and Nickolas S. Bloom — [etc.]
  1. Adsorption—Congresses. 2. Solution (Chemistry)—Congresses. I. Tewari, P. H. II.
American Chemical Society. Division of Colloid and Surface Chemistry.
QD547.A38   541.3'453   81-10708
ISBN-13: 978-1-4613-3266-4   e-ISBN-13: 978-1-4613-3264-0   AACR2
DOI: 10.1007/978-1-4613-3264-0

Proceedings of a symposium on Adsorption from Aqueous Solutions
held March 24-27, 1980, as a satellite symposium to the
meeting of the American Chemical Society Division of Colloid
and Surface Chemistry, Houston, Texas

© 1981 Plenum Press, New York
Softcover reprint of the hardcover 1st edition 1981

A Division of Plenum Publishing Corporation
233 Spring Street, New York, N.Y. 10013

PREFACE

Adsorption from aqueous solutions is important in many technological areas, like water purification, mineral beneficiation,
soil conservation, detergency, and many areas of biology. Recently,
adsorption of radionuclides from aqueous solutions has become the
focus of attention in assessing the movement of radionuclides
through a geologic medium from underground radioactive waste
repositories.

This volume provides a multidisciplinary overview of current
work in the area of adsorption from aqueous solutions, and reviews
the progress that has been made in the theoretical models for
assessing adsorption. Adsorption of heavy metal ions and the
effect of complex formation is treated extensively, as are the
effects of surface chemical properties of the adsorbent, solution
pH, and thermodynamic parameters important in the adsorption
process.

Adsorption of pesticides and organic polymeric species on
different adsorbents are included and implications of adsorption
of ions on dental materials are discussed. Also included are
studies of the adsorption of radionuclides by geologic media under
environmental conditions. The study of the chemical nature of the
adsorbed species at the surface by X-ray photoelectron spectroscopy which often provides mechanistic information for the
adsorption process is included for adsorbed metal ions on clay
and mineral surfaces.

I would like to express my thanks to the management of The
Research Company of the Atomic Energy of Canada, Ltd., Pinawa,
Manitoba, Canada, for permitting me to organize the symposium
and to edit the symposium volume. Thanks are also due to
Mr. I. Jenks of the Technical Services of the Atomic Energy of
Canada for providing technical and secretarial services. I would
also like to thank the reviewers of the papers for their time,
efforts, and many invaluable comments, and the authors for their

cooperation and interest. My special thanks are due to my wife, Shanti, for her understanding and cooperation during the editing, which happened to coincide with our move from Atomic Energy of Canada to Exxon Enterprises, Sunnyvale, California.

September 1981   Param H. Tewari
Sunnyvale, California

CONTENTS

THE POWER EXCHANGE FUNCTION: A GENERAL MODEL FOR

METAL ADSORPTION ONTO GEOLOGICAL MATERIALS

Donald Langmuir

Department of Chemistry and Geochemistry
Colorado School of Mines
Golden, Colorado 80401

ABSTRACT

The empirical data on adsorption of metal cations by naturally occurring soil materials can be systematized in terms of power exchange functions. For example, given the reaction $PbX + Ca^{2+} = CaX + Pb^{2+}$, the exchange function is: $K_{ex} = ([Pb^{2+}]/[Ca^{2+}])(CaX/PbX)^n$ where K_{ex} and n are constants, the brackets denote activities of the ions, and CaX and PbX are their mole fractions on sorbent X. This approach is mathematically equivalent to regular solution exchange when the mole fractions of two competing cations sorbed lie between 0.25 and 0.75, and to Freundlich isotherm-type behavior when the mole fraction of the minor cation is < 0.05. Combined literature review and laboratory study show that the exchange behavior of H^+, Na^+, K^+, Ca^{2+}, Mg^{2+}, Cd^{2+}, Co^{2+}, Ni^{2+}, Pb^{2+}, UO_2^{2+} and Zn^{2+} and their hydroxy-complexes on a variety of adsorbents (montmorillonites, beidellite, illite, ferric oxyhydroxides, zeolites, soils and humic materials) can be accurately described for a wide range of competing sorbate concentrations and ratios using from one to three power exchange expressions. The adsorption of the alkali metal and alkaline earth cations on pure clays at about 10^{-2} to 10^{-4} M often follows the power exchange function with $n = 1$, corresponding to simple ion exchange. Adsorption of heavy metals (between 10^{-3} and 10^{-7} M) is usually more complex, and fits power exchange functions with $n = 0.8$ to 2.0. Log-linearization of power exchange expressions yields lines with correlation coefficients usually in the range from 0.98 to 1.00. The power exchange model is a potentially useful predictor of heavy metal levels in the subsurface controlled by adsorption processes.

OVERVIEW OF METAL-ADSORPTION MODELS

     Geochemists and soil chemists attempting to explain and
predict metal adsorption in surface sediments, soil and groundwater systems, have available a wide variety of theoretical and
empirical adsorption models with which to approach the problem.
These include: (1) simple isotherm equations such as the Freundlich and Langmuir equations; (2) mass action type equations,
including Donnan exchange, and more general ion exchange expressions; and (3) composite models such as are presented by James
and Parks[1] and in this volume by James.[2] These models combine
isotherm and/or mass action type sorption concepts with double or
triple layer mathematical treatments, and expressions to define
speciation at the sorbent surface, and in solution.

     The mathematical simplicity of the isotherm and simple mass
action-type approaches is appealing. However, they are often poor
predictors of natural behavior. In part this is because they
usually assume a constant surface charge for sorbent materials,
and in the case of the isotherms, ignore competitive exchange
effects as well as other properties of the solution and sorbent.
On the other hand, the composite models, although they have shown
excellent success in explaining complex laboratory sorption
studies, and have a sound theoretical basis,[1-3] will be difficult
to apply to natural systems. In large part this is because of
the numerous parameters including intrinsic ionization and complexation constants for surface sites, and double layer capacitances, which must be known for each sorbent phase to make such
models work.

MODEL BACKGROUND

     This paper discusses a modified mass action exchange function
which we have found useful in the modeling of metal adsorption
behavior for a wide range of concentrations and activities of
competing ions, for variable pH values and variable surface
charge and at different temperatures. Our approach involves
computer calculation of solute activities using a greatly expanded
and modified version of the program WATEQ.[4] Model definition
also requires knowledge of the cation exchange capacity (CEC)
and/or base exchange capacity (BEC) of the sorbent material, as
well as the mole fractions of species adsorbed. The function for
binary exchange contains only two empirically determined constants.
Thus, for the homovalent binary exchange reaction,

     A + BX = B + AX   (1)

the empirical exchange constant expression is

$$K_{ex} = \frac{[B]}{[A]} \left(\frac{AX}{BX}\right)^n \tag{2}$$

where the square brackets denote molal activities of the enclosed solute species, AX and BX are mole fractions of A and B on the sorbent, n is constant, and K_{ex} is the exchange constant of the reaction.

The corresponding theoretical expression is

$$K_{ex} = \frac{[B]}{[A]} \frac{\lambda_{AX} (AX)}{\lambda_{BX} (BX)} \tag{3}$$

in which λ_{AX} and λ_{BX} are the respective rational activity coefficients on the sorbent surface. The Gibbs–Duhem equation for the binary exchange reaction is

$$(AX) \ d \ln \lambda_{AX} + (BX) \ d \ln \lambda_{BX} = 0 \tag{4}$$

Dividing by d(BX) and transposing, we find

$$d \ln \lambda_{AX}/d(BX) = -(BX \cdot d \ln \lambda_{BX})/AX \cdot d(BX) \tag{5}$$

Combining expressions (2) and (3) gives

$$\ln(\lambda_{AX}/\lambda_{BX}) = (n-1) \ln[(AX)/(BX)] \tag{6}$$

Taking the derivative of (6) with respect to BX leads to

$$d \ln \lambda_{AX}/d(BX) = (1-n)/(AX)(BX) + d \ln \lambda_{BX}/d(BX) \tag{7}$$

Eliminating $d \ln \lambda_{AX}/d(BX)$ between Equations (5) and (7), and simplifying yields

$$d \ln \lambda_{BX} = (n-1) \ d \ln(BX) \tag{8}$$

Integrating (8) between (BX) = λ_{BX} = 1 and non-unity values of (BX) and λ_{BX} gives

$$\lambda_{BX} = (BX)^{n-1} \tag{9}$$

A similar approach to λ_{AX} leads to

$$\lambda_{AX} = (AX)^{n-1} \tag{10}$$

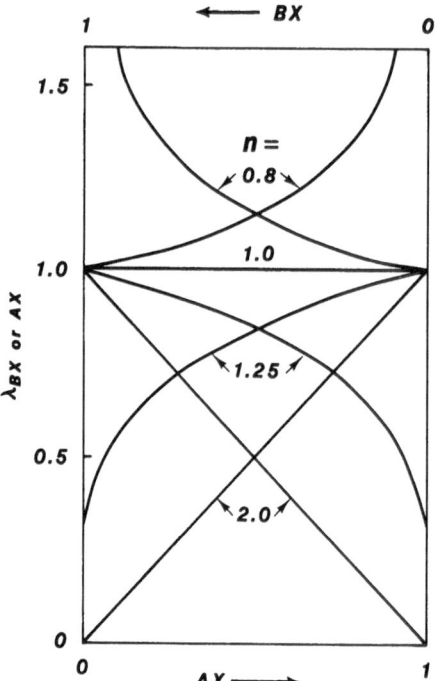

FIGURE 1   Plot of the rational activity coefficients for homo-
           valent binary exchange assuming power exchange functions
           with n = 0.8, 1.0, 1.25, and 2.00.

A plot of the rational activity coefficients versus mole
fractions adsorbed for the binary exchange, assuming four values
of n between 0.8 and 2.0, is given in Figure 1. This range of n
values includes nearly all study results we have fit to the
model. The figure shows that Raoult's Law is obeyed for the
major species adsorbed, and that for a given exchange constant,
when n is less than unity, the activity coefficient for the minor
component may greatly exceed unity, such that sorption of the
major component is preferred. Conversely, when n is greater than
unity, adsorption of the minor component is increasingly preferred
as the concentration of that component approaches zero. This
latter condition is the more typical of trace metal adsorption
behavior in natural systems. Figure 1 also shows that if (AX) is
equal to or less than 5% of (BX), and [B] nearly constant, then
to a good approximation $\lambda_{BX} \approx$ (BX) $\approx$ 1, and the power exchange
function reduces to the Freundlich isotherm equation

$$(AX) \ = \ K \ [A]^{1/n} \tag{11}$$

At trace concentrations of species A, the Langmuir isotherm
equation is mathematically equivalent to the Freundlich equation

when n = 1. Thus for ion A, the Langmuir isotherm equation may be written

$$N_{AX} = N_{max} \; b \; C/(1 + bC) \tag{12}$$

in which N_{AX} is the weight of A adsorbed per weight of sorbent, N_{max} the maximum weight of A that can be sorbed per weight of sorbent, b is a constant, and C the concentration of A in solution. For trace concentrations of A, expression (12) reduces to

$$N_{AX} = N_{max} \; b \; C \tag{13}$$

which, except for the units, is equivalent to expression (11) with n = 1. Expression (13) corresponds to Henry's Law behavior for the trace constituent, and can be thought of as the equation of the tangent to an adsorption isotherm plot drawn at (AX) = [A] = 0.

The power exchange function in expression (2) is identical to the so-called Rothmund-Kornfield Equation (c.f. Harmsen),[5] except that we are using free-ion activities rather than concentrations in solution. The function is also mathematically equivalent to a regular solution exchange model[6,7] when the mole fraction value of (AX) or (BX) lies between 0.25 and 0.75 (in this range the assumption that the power exchange function corresponds to regular solution behavior is valid to within 90% confidence limits). (See the discussion of regular solution models). However, in heavy metal adsorption studies comparable to natural conditions, the metal usually occupies much less than 25% of adsorption sites on a soil or individual sorbent phase relative to hydrogen ions or to the major cation species. We will see that the power exchange function model fits empirical exchange measurements involving both major and minor metals and protons over a wide range of mole fraction ratios, from unity up to two and three orders of magnitude.

Log linearization of the n-function for A - B exchange gives

$$\log \frac{[B]}{[A]} = \log K_{ex} + n \log \left(\frac{BX}{AX}\right) \tag{14}$$

so that a plot of log [B]/[A] versus log (BX)/(AX) has a slope of n, and intercept of log K_{ex}. Applied to published and unpublished studies of binary exchange at both constant and variable surface charge of sorbent, such plots are highly linear, with correlation coefficients (r values) usually between 0.98 and 1.00. Most published data complete enough to test the model, have been measured on clays and whole soils. The limited data available for sorption of lead and uranyl species versus hydrogen ion onto ferric oxyhydroxide also fits the model.

ADSORPTION ONTO WHOLE SOILS AND SEDIMENTS

Most published studies of metal adsorption onto whole soils
and sediments have dealt with the binary exchange behavior of
such materials with respect to H^+, Na^+, K^+, Ca^{2+}, and/or Mg^{2+}.
Levy and Hillel[8] examined Na^+ - Ca^{2+} exchange onto a silty-clay,
a clay-loam, and a sandy-loam soil. Their results for the clay-
loam soil are plotted in Figure 2 according to Equation (14).
Similar plots for the silty-clay and sandy-loam soils give K_{ex} =
1.12, n = 1.00 (r = 1.00); and K_{ex} = 2.37, n = 1.05 (r = 0.99),
respectively. For all three cases the empirical data plot above
a Donnan exchange line (drawn in this and following figures for
K_{ex} = 1, and n = 1). This indicates that Ca^{2+} adsorption is
favored over that of Na^+. The near perfect linearity of the
plots spanning approximately three orders of magnitude in the
$[Na^+]^2/[Ca^{2+}]$ ratio is considered remarkable.

Similar plots showing excellent fits to power exchange
functions are obtained when the data of Vanselow[9] for K^+ - Na^+
exchange onto "soil colloids" are plotted as in Figure 2.
Kittrick[10] has found that the experimental data for Ca^{2+} - Na^+
exchange onto 198 soils from six countries fits such a function
(r = 0.98) for over four orders of magnitude in the log equivalent
ratio of Na^+ to Ca^{2+}. Analysis of Ca^{2+} - Mg^{2+} exchange from
river water and sea water onto Amazon River suspended sediments
by Sayles and Mangelsdorf[11] showed that a good fit of the data
was obtained using a power exchange function. When their function
for the reaction Ca^{2+} + MgX = Mg^{2+} + CaX is written in the form
of Equation (2), the values of K_{ex} and n are 2.49 and 1.32,
respectively. The authors note that unpublished measurements of
Ca^{2+} - Mg^{2+} exchange on Dutch soils by van der Molen from fresh
waters and sea water fit a power exchange function in good agreement with their own.

Several authors have used the Freundlich equation to fit
trace heavy metal adsorption onto stream sediments. These include
Gardiner[12] who considered adsorption of trace concentrations of
Cd^{2+}, Cu^{2+}, and Hg^{2+} and O'Connor and Renn[13] who examined the
adsorption of trace zinc concentrations.

Bruggenwert and Kamphorst[14] have summarized the results of
numerous published trace heavy metal adsorption studies onto
soils in terms of a modified Langmuir isotherm equation, which is
equivalent to the Freundlich equation with n = 1 (see above).
The metals considered were divalent Cd, Co, Cu, Pb, and Zn.

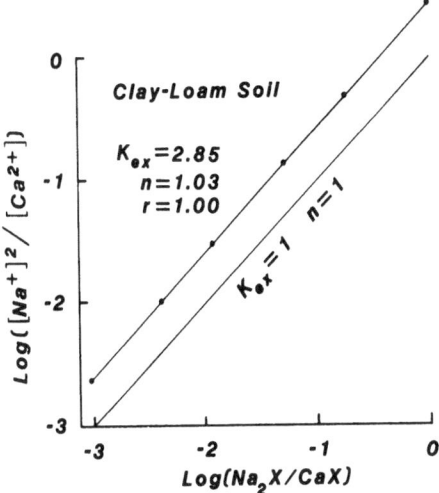

FIGURE 2   Power exchange function plot of Na^+ - Ca^{2+} exchange
onto clay-loam soil based on data from Levy and Hillel
(Reference (8)). "r" is the correlation coefficient
for the line through the data points. The line without
data points in this and subsequent plots (denoted here
by K_{ex} = 1 and n = 1) describes Donnan exchange behavior.

ADSORPTION ONTO ORGANIC MATTER

        The dominant role of recent organic matter in surface water
muds and some soils as a sorbent for trace heavy metals is well
known.[15-17] Published data to test the applicability of either
Freundlich or power exchange models to such adsorption is practically nonexistent, however. Frizado[18] has shown that a regular
solution model fits his empirical data for Na^+, K^+, Ca^{2+} and Mg^{2+}
versus H^+ exchange by humic materials. The fit is particularly
good at mole fractions of the metals adsorbed between 0.2 and
0.8, suggesting that Frizado's data would be at least as well
modeled by power exchange functions.

ADSORPTION ON CLAY MINERALS

        Extensive published data are available for testing the
applicability of the power exchange model to metal-clay adsorption (c.f. Oszvath,[19]; and references cited by Bruggenwert and
Kamphorst[14]). As might be expected, binary exchange on clay
minerals between Na^+ or K^+ and H^+, and between the alkaline
earths and alkali metals often fits power exchange functions with
n = 1.00 corresponding to simple ion exchange behavior.[19] In

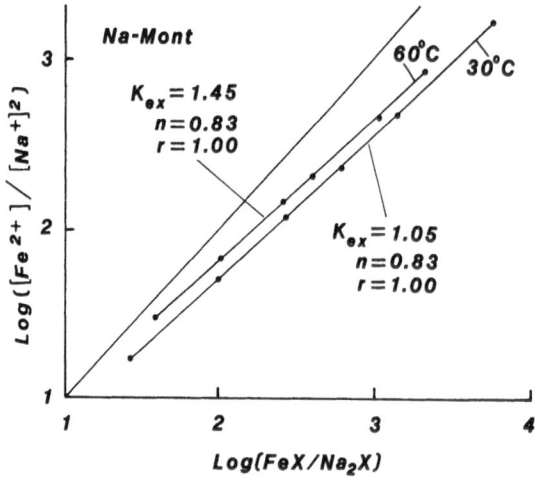

FIGURE 3   Power exchange function plot for Fe^{2+} - Na^+ exchange
           onto Na-montmorillonite at 30 and 60°C based on data
           from Singhal et al. (Reference (20)). "r" is the
           correlation coefficient for each line. The line with-
           out data points denotes Donnan exchange behavior.

most cases n values are also near unity for exchange onto clays
between the alkaline earths or alkali metals and species includ-
ing Cd^{2+}, Co^{2+}, Cu^{2+} and Zn^{2+} when the competing ions are approximately equal in aqueous concentrations. However, as a rule, when
the concentration of one of the competing metals is orders of
magnitude less than that of the other, adsorption of the minor
metal is preferred over the major.

Singhal et al.[20] measured Fe^{2+} - Na^+ exchange onto sodium
montmorillonite at 30 and 60°C, with Fe^{2+} and Na^+ concentrations
between 10^{-3} and 10^{-2} M. Their results plotted in Figure 3 show
that Fe^{2+} adsorption is preferred over that of Na^+ at both temperatures, but less so at 60°C. Application of the van't Hoff
equation to these data leads to an exchange enthalpy of 1.88
kcal/mol. Other studies of binary metal ion exchange as a function of temperature also fit power exchange equations (e.g.
Gaines and Thomas[21] (Ca^{2+} - Sr^{2+} onto montmorillonite, 5, 25, 50,
and 70°C); Van Bladel and Menzel[22] (Na^+ - Ca^{2+} onto bentonite,
25 and 50°C); El-Sayed and Burau[23] (Cu^{2+} - Ca^{2+} onto bentonite,
25 and 50°C); and Udo[24] (Ca^{2+} - Mg^{2+} onto kaolinite, 10 and
30°C)).

Bittel and Miller[25] examined Ca^{2+} - Cd^{2+} and Ca^{2+} - Pb^{2+}
exchange onto calcium montmorillonite. Their results for Ca^{2+} -
Cd^{2+} exchange are plotted in Figure 4 which shows Donnan-type
exchange behavior when $[Cd^{2+}] > [Ca^{2+}]$. A second function shows

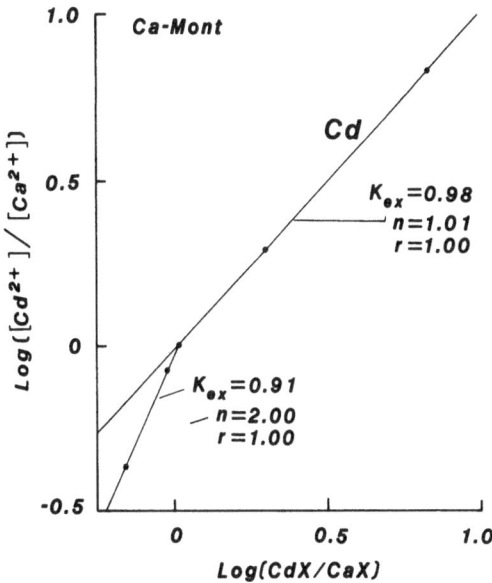

FIGURE 4   Power exchange function plot for Ca^{2+} - Cd^{2+} exchange
onto Ca-montomorillonite based on data from Bittel and
Miller (Reference (25)). "r" values are correlation
coefficients. The line without the points denotes
Donnan exchange behavior.

increasing selectivity of the clay for Cd^{2+} over Ca^{2+} with
decreasing Cd^{2+} concentrations. Bittel and Miller[25] did not
study the exchange reaction at Cd^{2+} levels low enough to typify
natural conditions. However, Ozsvath[19] has considered the same
exchange reaction at Cd^{2+} concentrations down to 0.1 mg/L ($\approx 10^{-6}$ M)
in the presence of about 10^{-3} M Ca^{2+} (see Appendix). His results
are shown in Figure 5. The upper segment of his plot is co-
linear with the non-Donnan portion of Bittel and Miller's plot in
Figure 4. Both segments of Figure 5 demonstrate increased pref-
erence of the clay for Cd^{2+} over Ca^{2+} as $[Cd^{2+}]/[Ca^{2+}]$ decreases.
When considered together Figures 4 and 5 indicate three separate
power exchange functions will fit the data for Ca^{2+} - Cd^{2+}
exchange over nearly four orders of magnitude in the aquo-ion
ratio. That each segment may correspond to a different sorption
mechanism is possible but somewhat conjectural.

Bittel and Miller[25] also considered Ca^{2+} - Pb^{2+} exchange
onto calcium montmorillonite. Their results plotted in Figure 6
show that Pb^{2+} sorption is favored over that of Ca^{2+} at all
concentration ratios, but increases as $[Pb^{2+}]/[Ca^{2+}]$ decreases.
Ozsvath's experimental results[19] for the same exchange reaction
measured for geochemically realistic concentrations of Pb^{2+}

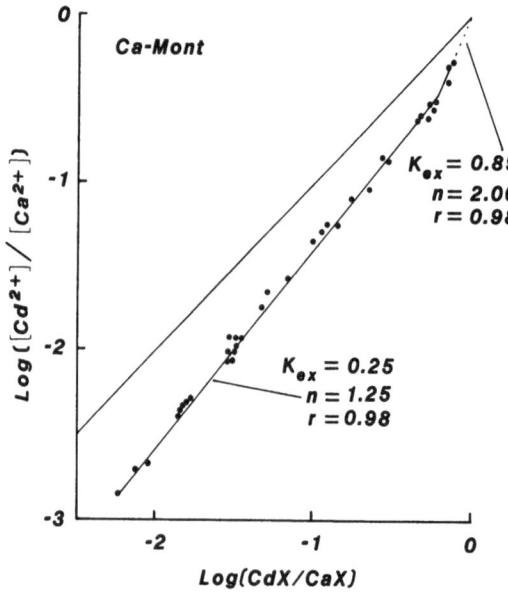

FIGURE 5   Power exchange function plot for Ca^{2+} - Cd^{2+} exchange
onto Ca-montomorillonite based on data in the Appendix
from Ozsvath (Reference (19)).   The solid line denotes
Donnan exchange behavior.

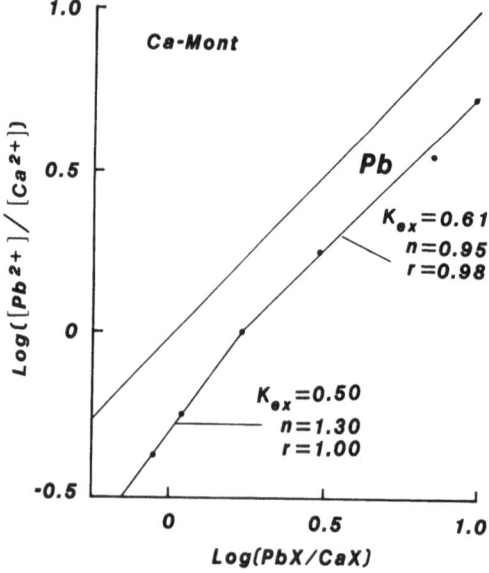

FIGURE 6   Power exchange function plot for Ca^{2+} - Pb^{2+} exchange
onto Ca-montmorillonite based on data from Bittel and
Miller (Reference (25)).   The solid line denotes Donnan
exchange behavior.

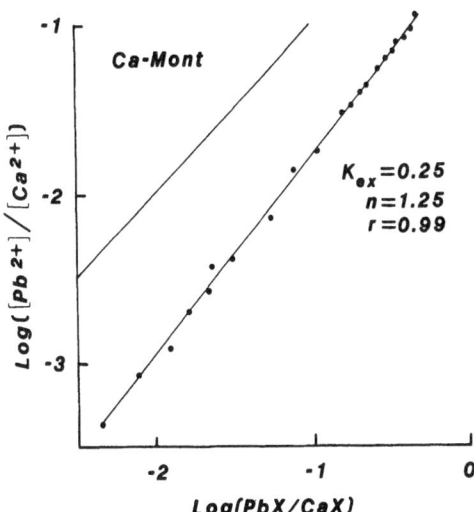

FIGURE 7   Power exchange function plot for Ca^{2+} - Pb^{2+} exchange
onto Ca-montomorillonite based on data in the Appendix
from Ozsvath (Reference (19)).   The solid line denotes
Donnan exchange behavior.

(down to 3.4×10^{-7} M) and Ca^{2+} (at about 10^{-3} M) are given in
the Appendix and plotted in Figure 7.   The slight disagreement
between the lower line in the Bittel and Miller plot and the line
in Figure 7 may reflect the fact that Bittel and Miller equilibrated their adsorption runs for four hours, whereas Ozsvath's
were equilibrated for 24 hours.   In Ozsvath's study, solution
activities of Cd^{2+} and Pb^{2+} were corrected for $CdOH^+$ and $PbOH^+$
complexing.   The Ca^{2+} - Cd^{2+} data were measured at pHs between
5.2 and 8.5, the Ca^{2+} - Pb^{2+} data between pH 5.2 and 5.7.   Mole
fractions adsorbed were computed taking into account the known pH
variations of the base exchange capacity (cation exchange capacity
minus proton exchange capacity).

ADSORPTION ONTO FERRIC OXYHYDROXIDES

    Although an extensive literature search proves the importance
of manganese oxide adsorption of trace heavy metals in soils and
sediments,[26,27] there is a surprising paucity of adequate empirical data describing such adsorption.   The published data are
generally incomplete in ways that preclude casting it into the
form of a power exchange function.

    Limited data for metal ion $-H^+$ exchange onto ferric oxy-
hydroxides are available, and they fit the power exchange function.

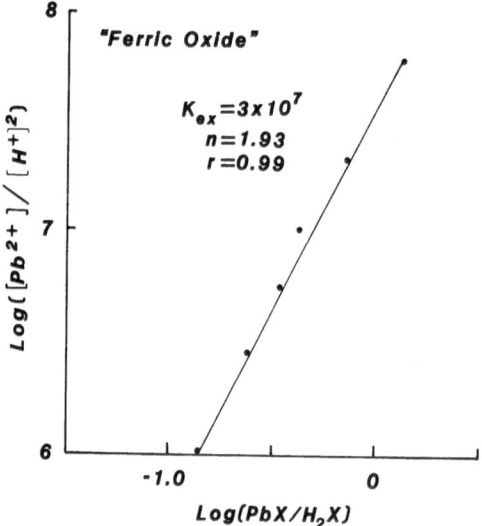

FIGURE 8   Power exchange function plot for Pb^{2+} - H^+ exchange
onto "ferric oxide" based on data given by Gadde and
Laitinen (Reference (28)).

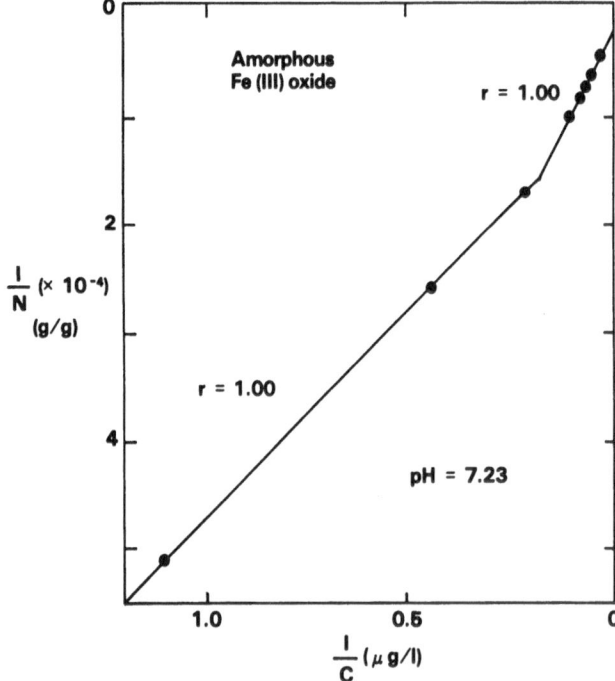

FIGURE 9   Linearized Langmuir isotherm plot of uranyl adsorption
onto amorphous ferric oxyhydroxide.

Gadde and Laitinen[28] considered Pb^{2+} - H^+ exchange on "ferric oxide" at pH 5. Their results are plotted in Figure 8. The Donnan line (not shown) would plot off the graph to the lower right, indicating that proton adsorption is greatly preferred over that of Pb^{2+}.

A linearized Langmuir isotherm plot of uranyl -H^+ exchange onto amorphous ferric oxyhydroxide at pH 7.23 is shown in Figure 9. The data used to generate the plot are from van der Weijden and Langmuir,[29] and Hsi and Langmuir.[30] Because the plot exhibits two linear segments, the conventional interpretation would be that each segment corresponds to a different adsorption mechanism. The same data are cast in power exchange form in Figure 10. Computer calculation of uranyl speciation in solution using an expanded version of the program WATEQF,[4] with thermochemical data for uranium species from Langmuir,[31] shows that practically all (> 99%) of the uranyl is present as UO_2OH^+ ion. If we assume this is the adsorbed species as well, a single function fits all the data. Regardless of the adsorption mechanism or mechanisms, Figure 10 shows that adsorption of UO_2OH^+ is increasingly favored over that of protons at total uranyl concentrations down to $10^{-9.6}$ M (0.056 μg/L).

COMPARISON OF THE POWER EXCHANGE AND REGULAR SOLUTION MODELS

For the binary exchange reaction $B^+ + AX = A^+ + BX$, the exact equation for the exchange according to regular solution theory is

$$\log K'_{ex} = [\log K_{ex} + \omega/2.303\ RT] - [2\omega\ (AX)/2.303\ RT] \quad (15)$$

In this expression $K'_{ex} = [A^+](BX)/[B^+](AX)$, K_{ex} is the thermo-dynamic exchange constant, ω is constant at constant temperature and pressure, and denotes the energy of interaction between A and B at the sorbent surface.[7] Garrels and Christ[6] and Truesdell and Christ[7] show that λ_{AX} and λ_{BX} are related to ω through the equations

$$\lambda_{AX} = \exp\ [\omega\ (AX)^2/RT] \quad (16)$$

and

$$\lambda_{BX} = \exp\ [\omega\ (BX)^2/RT] \quad (17)$$

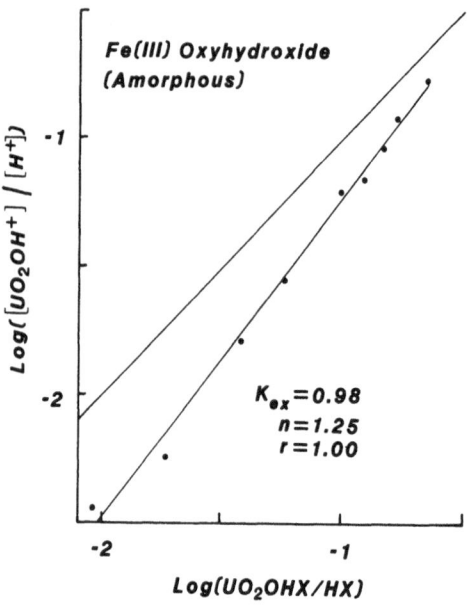

FIGURE 10   Power exchange function plot of the data shown in
                 Figure 9. The line without data points denotes
                 Donnan exchange behavior.

The form of Equation (15) is such that, if regular solution
theory applies, a plot of log K'_{ex} versus (AX) should yield a
straight line with a slope of $-(2\omega/2.303\ RT)$ and intercept equal
to (log K_{ex} + $\omega/2.303\ RT$).

     Truesdell and Christ[7] plot log K'_{ex} versus (HX) computed from
the empirical data for K^+ - H^+ exchange onto H-biedellite published
by Marshall and Bergman.[32] This plot, given in Figure 4 of the
paper by Truesdell and Christ, is in the form of a parabolic
curve rather than appearing as one or more straight lines. The
authors argue that the exchange reaction can be explained by
assuming regular solution behavior and two exchange sites. The
same empirical data are plotted in the form of a power exchange
function in Figure 11. The perfect fit of the data to two straight
line segments indicates that the exchange reaction is better
described by power exchange functions than by assuming regular
solution behavior.

CONCLUSIONS

     The power exchange function is a simple two-parameter equa-
tion that very accurately models the behavior of many major and
minor cation exchange reactions. It is mathematically equivalent
to the Freundlich isotherm equation at low and high mole fraction

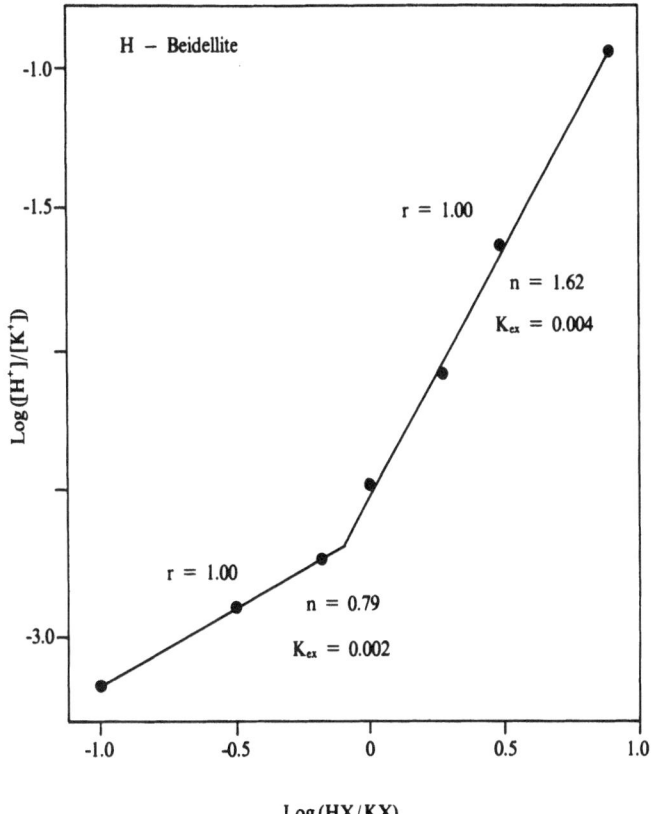

FIGURE 11   Power exchange function plot for H^+ - K^+ exchange
               onto H-biedellite based on the data of Marshall
               and Bergman (Reference (32)).

adsorbed ratios, and to regular solution exchange at intermediate
mole fraction ratios. More research is needed to explain its
remarkable success on a sound theoretical basis. Also, more
empirical adsorption data are needed, so that we can better
generalize and predict exchange constants and n values for a wide
variety of competing ions and sorbents and at different temperatures for application to metal adsorption in natural systems.

ACKNOWLEDGEMENTS

        This material is based upon work supported by the Office of
Surface Mining, Department of the Interior, under Grant No.
G51915011. Any opinions, findings, and conclusions or recommen-
dations expressed in this publication are those of the author and
do not necessarily reflect the views of the Office of Surface
Mining, Department of the Interior. Many graduate students have

contributed generously to the analyses of adsorption data discussed in this paper. These include David Ozsvath at the Pennsylvania State University and John Catts and Daniel Hsi at the Colorado School of Mines.

REFERENCES

1. R.O. James and G.A. Parks, Characterization of aqueous colloids by their electrical double layer and intrinsic surface chemical properties, in: "Surface and Colloid Sci.", Vol. 12, E. Matijevic, ed., Plenum Press, New York (in press).
2. R.O. James, The TiO_2/aqueous electrolyte system - applications of colloid models and model colloids, in: "Adsorption from Aqueous Solutions", P.H. Tewari, ed., Plenum Press, New York (1981).
3. J.A. Davis and J.O. Leckie, Speciation of adsorbed ions at the oxide/water interface, in: "Chemical Modeling in Aqueous Systems", E.A. Jenne, ed., Am. Chem. Soc. Symp. Ser. 93:299-317 (1979).
4. L.N. Plummer, B.F. Jones and A.H. Truesdell, WATEQF - A FORTRAN IV version of WATEQ, a computer program for calculating chemical equilibrium of natural waters, U.S. Gel. Surv., Water-Resources Inv. 76-13.
5. K. Harmsen, Theories of cation adsorption by soil constituents: Discrete-site models, in: "Soil Chemistry. B. Physico-Chemical Models", G.H. Bolt, ed., Elsevier Scientific Publ. Co., New York (1979). pp. 77-139.
6. R.M. Garrels and C.L. Christ, "Solutions, Minerals, and Equilibria", Freeman, Cooper and Company, San Francisco, CA. (1965).
7. A.J. Truesdell and C.L. Christ, Am. J. Sci. 266:402 (1968).
8. R. Levy and D. Hillel, Soil Sci. 106(5):393 (1968).
9. A.P. Vanselow, Soil Sci. 33:95 (1932).
10. J.A. Kittrick, Soil Sci. Am. Jour. 40:147 (1976).
11. F.L. Sayles and P.C. Mangelsdorf, Jr., Geochim. Cosmochim. Acta 43:767 (1979).
12. J. Gardiner, Water Research 8:157 (1974).
13. J.T. O'Connor and C.E. Renn, Am. Water Works Assoc. 56(8):1055 (1964).
14. M.G.M. Bruggenwert and A. Kamphorst, Survey of experimental information on cation exchange in soil systems, in: "Soil Chemistry. B. Physico-Chemical Models", G.H. Bolt, ed., Elsevier Scientific Publ. Co., New York (1979).
15. S.J. Toth and A. Ott, Envir. Sci. Technol. 14:935 (1970).

16. E.A. Jenne and S.N. Luoma, The forms of trace elements in soils, sediment, and associated waters, in: "Biological Implications of Metals in the Environment", R.E. Wildung and H. Drucker, eds., CONF-750929, NTIS, Springfield, VA. (1977). pp. 110-143.

17. G.H. Bolt, "Soil Chemistry. B. Physico-Chemical Models", Elsevier Scientific Publ. Co., New York (1979). pp. 479-.

18. J.P. Frizado, Ion exchange on humic materials - a regular solution approach, in: "Chemical Modeling in Aqueous Systems", E.A. Jenne, ed., Am. Chem. Soc. Symp. Series 93 (1979). pp. 133-145.

19. D. Ozsvath, "Modeling Heavy Metal Sorption from Subsurface Waters with the n-Power Exchange Function". M.S. Thesis in Geochemistry, Pennsylvania State University, University Park, PA. (1979). pp. 61-.

20. J.P. Singhal, D. Kumar and G.K. Gupta, Ind. J. Chem. 14A:929 (1976).

21. G.L. Gaines and H.D. Thomas, J. Chem. Phys. 23:2322 (1955).

22. R. Van Bladel and R. Menzel, "A thermodynamic study of sodium-strontium exchange on Wyoming bentonite", Internatl. Clay Conf., 1969, pp. 619-634.

23. M.H. El-Sayed and R.G. Burau, Soil Soc. Am. Proc. 34:397 (1970).

24. E.J. Udo, Soil. Sci. Am. Jour. 42:556 (1978).

25. J.E. Bittel and R.J. Miller, J. Envir. Quality 3:250 (1974).

26. D.L. Suarez and D. Langmuir, Geochim. Cosmochim. Acta 40(6):589 (1976).

27. E.A. Jenne, Trace element sorption by sediments and soils - sites and processes, in: "Symp. on Molybdenum in the Environment", W. Chappel and K. Petersen, eds., M. Dekker Inc., New York (1977). pp. 425-553.

28. R.R. Gadde and H.A. Laitinen, Envir. Letters 5:223 (1973).

29. C.H. van der Weijden and D. Langmuir, unpublished manuscript (1976).

30. C-K.D. Hsi and D. Langmuir, manuscript in preparation (1980).

31. D. Langmuir, Geochim. Cosmochim. Acta 42:589 (1978).

32. C.E. Marshall and W.E. Bergman, J. Phys. Chem. 46:52 (1942).

APPENDIX

Tables A1 and A2.   Experimental and calculated data for the
                     adsorption of Cd and Ca (Table A1), and Pb and
                     Ca (Table A2) on montmorillonite at 25°C.
                     $[Cd^{2+}]$, $[Pb^{2+}]$ and $[Ca^{2+}]$ are the equilibrium
                     (mol/L) activities of these species in solution
                     following adsorption. CdX, PbX, and CaX are
                     mole fractions of the metal ions adsorbed.
                     The pH is the equilibrium value, and BEC (in
                     meq/100 g) is the base exchange capacity of
                     the clay at that pH.   The data are from Refer-
                     ence (19).

Table A1

pH	BEC	-log $[Cd^{2+}]$	-log $[Ca^{2+}]$	-log (CdX/CaX)	pH	BEC	-log $[Cd^{2+}]$	-log $[Ca^{2+}]$	-log (CdX/CaX)
6.60	93	5.90	3.06	2.24	6.80	96	4.40	3.06	1.01
6.75	95	5.82	3.10	2.12	8.30	115	4.39	3.10	0.95
7.00	98	5.77	3.11	2.04	7.95	111	4.36	3.10	0.91
7.90	110	5.47	3.07	1.86	6.30	88	4.32	3.08	0.85
7.10	100	5.43	3.07	1.84	7.80	109	4.20	3.10	0.75
6.00	83	5.39	3.06	1.82	6.75	95	4.10	3.06	0.63
5.15	68	5.37	3.06	1.80	7.55	106	3.97	3.10	0.52
4.10	43	5.35	3.06	1.78	7.55	106	3.95	3.09	0.54
7.00	98	5.14	3.07	1.53	7.00	98	3.69	3.16	0.24
6.00	83	5.13	3.06	1.52	6.75	95	3.69	3.07	0.27
5.20	69	5.12	3.08	1.50	6.20	87	3.68	3.07	0.32
6.45	90	5.08	3.13	1.48	6.00	83	3.68	3.08	0.30
6.80	95	5.08	3.06	1.53	7.30	103	3.65	3.10	0.24
6.10	85	5.07	3.06	1.51	7.15	101	3.61	3.09	0.23
5.20	69	5.06	3.07	1.50	7.00	98	3.47	3.07	0.15
6.20	87	4.80	3.07	1.33	7.00	98	3.37	3.07	0.13
7.00	98	4.70	3.04	1.29	7.00	98	3.37	3.09	0.11
6.00	83	4.63	3.05	1.16	7.00	98	3.37	3.09	0.11
6.00	83	4.49	3.06	1.03					

Table A2

pH	BEC	-log $[Pb^{2+}]$	-log $[Ca^{2+}]$	-log (PbX/CaX)	pH	BEC	-log $[Pb^{2+}]$	-log $[Ca^{2+}]$	-log (PbX/CaX)
5.15	68	6.47	3.09	2.32	5.65	77	4.51	3.10	0.67
5.50	74	6.17	3.08	2.10	5.60	76	4.45	3.10	0.62
5.60	76	5.97	3.09	1.77	5.70	78	4.41	3.10	0.60
5.60	76	5.75	3.05	1.77	5.40	73	4.37	3.10	0.56
5.60	76	5.68	3.09	1.65	5.55	75	4.34	3.08	0.54
5.50	74	5.51	3.07	1.62	5.40	73	4.30	3.09	0.51
5.70	78	5.44	3.06	1.49	5.55	75	4.27	3.11	0.47
5.60	76	5.23	3.07	1.25	5.65	77	4.23	3.11	0.45
5.65	77	4.93	3.06	1.10	5.55	75	4.17	3.08	0.39
5.60	76	4.81	3.06	0.95	5.55	75	4.14	3.10	0.34
5.60	76	4.63	3.10	0.79	5.70	78	4.07	3.12	0.31
5.60	76	4.56	3.10	0.72					

THE TiO$_2$/AQUEOUS ELECTROLYTE SYSTEM - APPLICATIONS OF COLLOID MODELS AND MODEL COLLOIDS

R.O. James,[†], P.J. Stiglich* and T.W. Healy*

†CSIRO Physical Technology Unit
Institute of Earth Resources
338 Blaxland Road
Ryde, N.S.W. 2112 Australia

*Department of Physical Chemistry
University of Melbourne
Victoria 3052 Australia

In the past decade, there have been some considerable advances in the study of the colloid chemical properties of hydrosols. These advances resulted in part from the increased attention paid to (a) characterisation of the type and site density of reactive functional groups on the colloid, (b) the tendency of the functional groups to react with ionic species from the solution phase and (c) the use of electrical double layer (e.d.l.) models to estimate the effect of the development of surface charge and potential in the e.d.l. on the tendency of the functional groups to react. This type of approach has now been used for description of the potentiometric and conductometric titrations, the surface charge density and electrokinetic potential of polymer latices, oxides and clays dispersed in aqueous electrolytes. Such descriptions can also be extended to the adsorption of dilute solutes, e.g. hydrolysable metal ions such as cadmium onto hydrosols.

Together with this progress many novel model colloids have been reported although only few have had their adsorption properties studied in detail. These will provide future opportunities for improving colloid models.

INTRODUCTION

    The interactions of ions and molecules with the surfaces of
colloid particles dispersed in aqueous solutions are subject to
continuing investigation in industrial, agricultural and environmental sciences. This is partly due to interest in the removal
of ions and molecules from solution by adsorption, which one may
regard in some aspects as an effect exerted on the ions and
molecules by the solid surface. In addition, the role of ions in
controlling the surface charge and the electrophoretic mobility
or zeta potential of the colloidal particles is of great importance
in the rates of particle coagulation and adhesion to other surfaces. One may take the view that the colloid stability behavior
reflects effects exerted on the surface and its electrical double
layer by the adsorbed ions.

    Thus, for the reliable characterization of the colloid
chemical properties of hydrosols the most important properties
necessary are adsorption densities of ions, the surface charge
densities and electrokinetic potential.[1] However, it is surprising to find that very few papers in the literature investigate
all of these properties for any given system. Thus, for example,
in environmental and chemical modelling studies, where there is
obvious interest in the uptake of ionic species, it is often
found that many adsorption isotherms are reported, but there are
few or no determinations of the surface charge or zeta potential
of the system. Whilst such a study may be a necessary part of
the investigation of a problem, without the complementary surface
charge and potential data it offers little to a fundamental
understanding of the important reactions and processes in hydrosols.[1,2]

    One of the more difficult problems facing the colloid chemist
trying to interpret adsorption reactions and associated phenomena
is that more often than not he does not know the form or state of
the adsorbed entity and the adsorption site. Consequently, most
adsorption reactions used in colloid models are postulates that
are consistent with the stoichiometry of ions disappearing from
or appearing in the solution phase. New techniques, e.g., ESCA,
have been applied to determine the surface states at the high
vacuum/solid interface.[3,4,5,6] Whether these results reliably
reflect the situation at the aqueous electrolyte/solid interface
is not always clear, since evaporation of the aqueous phase
always increases ionic concentrations and the pH changes. Another
area lacking attention in the colloid chemistry of hydrosols is
the kinetic investigation of the mechanisms of adsorption in the
electrical double layer. Recent work by Ashida et al.[7,8] on
ionic adsorption-desorption at oxide/water interfaces by relaxation techniques shows considerable promise if the technique can
be extended to a wider range of electrolyte concentration.

However, until these newer techniques become more widely
available, much of our knowledge of hydrosol properties will
continue to rely upon equilibrium and steady state measurements
of ionic adsorption density, surface charge density and electrokinetic potential.

To complement the application of the newer techniques,
advances continue to be made in other areas such as computational
models for the reactions in the electrical double layer. The
approach in these models is to try to answer the following questions about the nature of colloid surfaces. What is the density
of surface functional groups, reaction sites or adsorption sites;
what is the tendency of these sites to react with ions from the
solution; and how does the development of the electrical double
layer affect the tendency of the sites to react with solutes?
These questions are similar to those posed by Scatchard[9] in his
discussion of the attraction of protein for small molecules and
ions, namely, how many, how tightly, where and why? Another area
in which advances have been made is in the preparation of model
colloids, that is, colloidal particles of uniform size, morphology
and surface structure. These include the many oxide hydrosols
prepared by Matijevic and associates[10-15] and also the organic
polymer latices prepared and characterised in a number of laboratories.[16-22] To date, relatively little use of these model
colloids has been made in efforts to test colloid models.[19,23]

This paper reports the investigation of adsorption reactions
at the titanium dioxide/water interface in solutions of background
electrolyte in the presence and absence of dilute concentrations
of cadmium nitrate. It is one of the few reports in which adsorption density, surface charge density and electrokinetic measurements have been made. The results are interpreted using surface
complexation model or site binding model described by Davis,
James and Leckie[23-25] and in a recent review.[1] The titania used
here although not fulfilling all the requirements of a model
colloid, is readily available, is of fairly uniform shape and
size, is readily purified, has very low solubility and the point
of zero charge (pH_{pzc}) and isoelectric point (pH_{iep}) are conveniently near neutral pH conditions.

EXPERIMENTAL

The oxides used in these studies were high purity research
samples of rutile (TiO_2), products of Tioxide International
(preparation numbers CLD528 and CLD640) which were prepared by
hydrolysis of titanium tetrachloride. To ensure removal of
chloride ions, the oxides were washed over a forty-eight hour
period by Soxhlet extraction,[26] then freeze dried and stored as
dry powder. Electron microscopy showed the particles to be

prolate ellipsoids of essentially uniform size. The specific
surface areas (BET, N_2 gas, 16.2 Å/molecule) determined by gravi-
metric adsorption were 19.8 m^2/g for CLD528[27] (Cahn R.G. Automatic
Electrobalance), and 15.4 m^2/g for CLD640.[26,28]

The adsorption of cadmium onto TiO_2 (CLD528, 10 g/dm^3) was
determined by observing the loss of cadmium hydroxide from a
solution by using both ion selective electrode and atomic absorption spectroscopy techniques as reported earlier.[26,28] In blank
2 x 10^{-4} mol dm^{-3} cadmium nitrate solutions, and 1 x 10^{-2} mol dm^{-3}
KNO_3 the response of the ion selective electrodes indicated that
no significant hydrolysis of Cd^{2+} occurred for pH values below 8
and that hydrolysis and precipitation of cadmium hyroxide occurred
for pH values greater than 9.[26,28]

Potentiometric titrations were performed in a closed vessel
under a humidified high purity N_2 atmosphere,[26,27,29] using
5 g dm^{-3} of TiO_2 (CLD640) and 1 x 10^{-4} mol dm^{-3} $Cd(NO_3)_2$. Water
used was triply distilled in an all glass apparatus. KNO_3 used
as background electrolyte was purified by repeated hot filtration,
recrystallization and recovery of analytical reagents. Buffers
used were Merck Titrisol with values 4.01, 6.98 and 8.95 at 25°C.

The net protonic charge density measured by potentiometric
titration is a function of pH and is given by

$$\sigma_H = F\ (\Gamma_{H^+} - \Gamma_{OH^-})\qquad\qquad C/cm^2$$

$$= F\ (C_A - C_B + [OH^-] - [H^+])/A$$

where Γ_i are adsorption densities (mol/cm^2), A is the surface
area of the suspension cm^2 dm^{-3} and C_A and C_B are the concentrations of acid or base after addition and F is the Faraday constant.[1,23,25] The acids and bases used (HNO_3 and KOH) were
prepared from BDH concentrated volumetric solutions.

The electrophoretic mobility of TiO_2 particles in various
electrolyte compositions was measured using a Rank Bros. particle
microelectrophoresis apparatus Mark II, with the rectangular
cell. Particles were timed successively using opposite applied
polarity to minimise concentration effects. Usually 10 particles
were timed in each direction also at the front and rear stationary
levels.[26]

RESULTS

Two rutile powders were used in the series of experiments.
The rutile used for the experiments in which the adsorption of

cadmium was measured by direct methods was code named CLD528.
The surface charge and electrokinetic potential of this sample
has been investigated previously by Yates[27,29] and Wiese,[30,31]
respectively.

     The rutile powder used for the surface charge measurements,
the potentiometric titration in the presence of cadmium solutions
and the electrokinetic potential measurements was code named
CLD640.

     The results of the potentiometric titration of TiO₂ in
aqueous KNO_3 solutions (CLD640) is shown in Figure 1. The characteristic increase in the magnitude of the surface charge with
increasing electrolyte concentration is observed. The intersection
of the adsorption isotherms gives the point of zero charge (pH_{pzc})
as 5.8. Also shown in the figure is the dependence of the zeta
potential of TiO₂ on pH at constant 10^{-3} M KNO_3. The isoelectric
point (pH_{iep}) is 5.8, coincident with the pH_{pzc}.

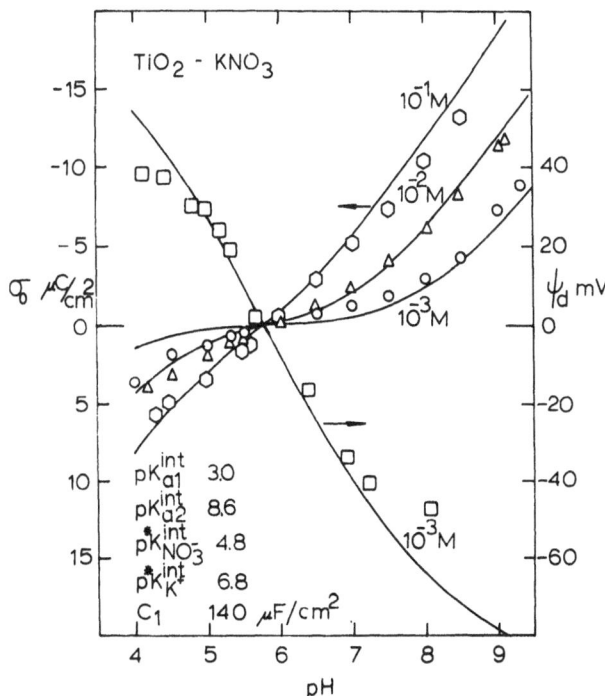

FIGURE 1   The surface charge density and zeta potential of the
           TiO₂ (Batch CLD640) - H₂O interface as a function of pH
           at various aqueous KNO_3 concentrations. Points are
           experimental and lines are calculated using the e.d.l.
           model described in the text.

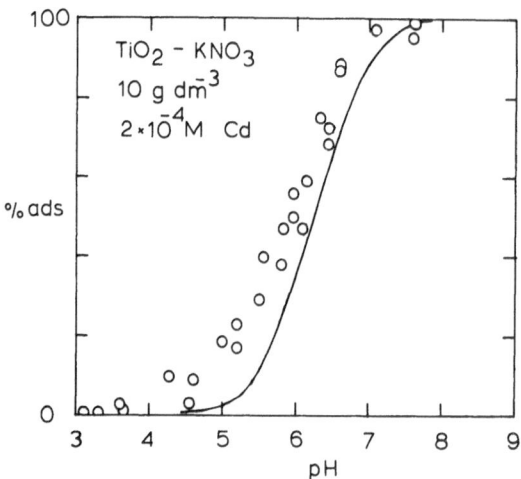

FIGURE 2   The percentage adsorption of cadmium by TiO_2 (CLD528),
10 g dm^{-3} from 2 x 10^{-4} mol dm^{-3} KNO_3 background electrolyte.   The line is calculated using the e.d.l. model
with surface chemical characteristics for TiO_2 (CLD528).

The percentage of cadmium removed from aqueous solution
(2 x 10^{-4} mol dm^{-3} $Cd(NO_3)_2$, 10^{-2} mol dm^{-3} KNO_3) by adsorption
onto rutile (CLD528, 10 g dm^{-2}, 19.8 m^2 g^{-1}) is shown as a function
of pH in Figure 2.   This data repeats the often observed pattern
for the uptake of cations by oxides, that is, increasing adsorption
with increasing pH.

Concurrent with adsorption, protons are released from the
interface.   This is shown in the potentiometric titration of the
TiO_2 (CLD640, 5 g dm^{-3}, 15.8 m^2/g) in aqueous cadmium solutions
(1 x 10^{-4} mol dm^{-3} $Cd[NO_3]_2$, 10^{-3} mol dm^{-3} KNO_3) in Figure 3.
Also shown are titrations for

1. 10^{-3} mol dm^{-3} KNO_3 solution;

2. 10^{-4} mol dm^{-3} Cd $(NO_3)_2$ in 10^{-3} mol dm^{-3} KNO_3, and

3. TiO_2 in 10^{-3} mol dm^{-3} KNO_3.

Differences between such curves allow estimation of adsorption of
H^+ or OH^- on TiO_2 in simple salt solutions, the extent of hydrolysis of Cd^{2+} and the release of H^+ or consumption of OH^- accompanying the adsorption of cadmium.

The protonic charge σ_H released as a consequence of cadmium
adsorption is shown in Figure 4.   Also shown is the estimate of

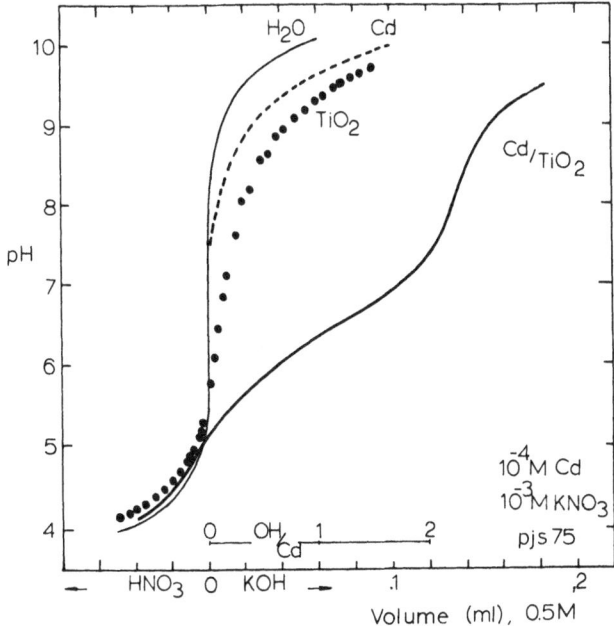

FIGURE 3   The potentiometric titration of various components in
the aqueous $Cd(NO_3)_2/KNO_3/TiO_2$ hydrosol. Line and
points are experimental. H_2O signifies the strong acid
(HNO_3) - strong base (KOH) titration in water: Cd
signifies titration of 10^{-4} mol dm^{-3} $Cd(NO_3)_2$ in
10^{-3} mol dm^{-3} KNO_3 with KOH; TiO_2 signifies the titration of TiO_2 in 10^{-3} mol dm^{-3} KNO_3 with strong acid
(HNO_3) and base (KOH) and Cd/TiO_2 signifies the titration of the mixed system 10^{-4} mol dm^{-3} $Cd(NO_3)$, 5.0 g/dm^3
TiO_2 and 10^{-3} mol dm^{-3} KNO_3 with strong acid and base.

the surface charge density of the TiO_2 derived from the release
of protons according to the protonic mass balance

$$\sigma_o = F(\Gamma_{SOH_2^+} + \Gamma_{SOH_2NO_3^+} - \Gamma_{SO^-} - \Gamma_{SO^-K^+} - \Gamma_{SO^-CdOH^+})\ C/cm^2$$

This will be discussed later.

The zeta potential of TiO_2 (CLD640, 0.05 g dm^{-3}, 10^{-3} mol dm^{-3}
KNO_3) derived from electrophoretic mobility measurements is shown
in Figure 5 for the absence of cadmium nitrate and then in the
presence of 10^{-5} and 10^{-4} mol dm^{-3} of cadmium nitrate. The
adsorption of cadmium changes the magnitude and sign of the
diffuse double layer potential and charge under some conditions.

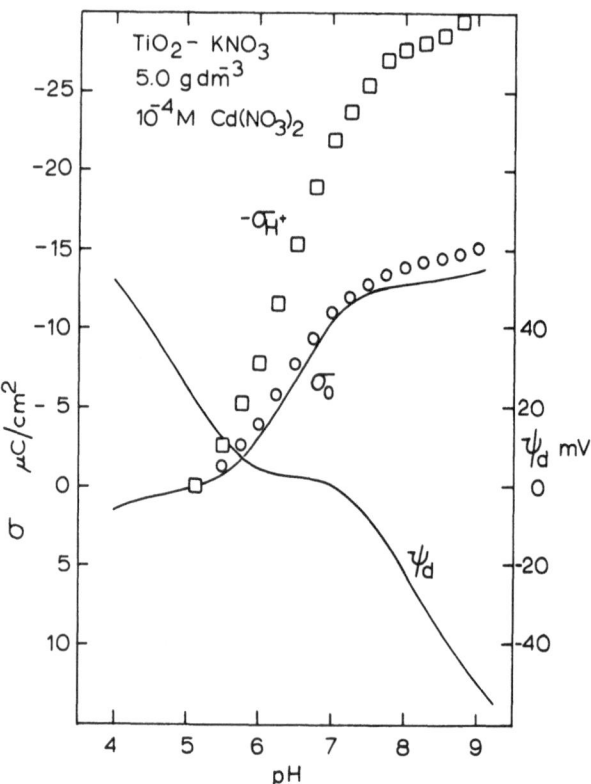

FIGURE 4   Net protonic charge σ_H released during adsorption of
cadmium on TiO_2 (CLD640) in aqueous 10^{-3} mol dm^{-3} KNO_3
solutions.   Also shown is the surface charge density σ_o
derived from σ_H assuming the adsorbed cadmium state is
represented by Equation (20) as $-SO^-CdOH^+$.   Lines are
calculated using e.d.l. model described in text with
surface chemical characteristics for TiO_2 CLD640.

DISCUSSION

A.   Application of Colloid Models to Experimental Results

      Various models used to describe the surface reactions and
the e.d.l. properties of oxide colloids in aqueous electrolytes
have been recently reviewed.[1,30,31]   One may consider the surface
hydroxyl groups formed by reaction of water with the oxide powder
as reactive amphoteric sites which may then dissociate or adsorb
ions.   In dilute background electrolyte solutions the dominant
reactions are the uptake or loss of protons at neutral surface
hydroxyls, SOH, e.g.

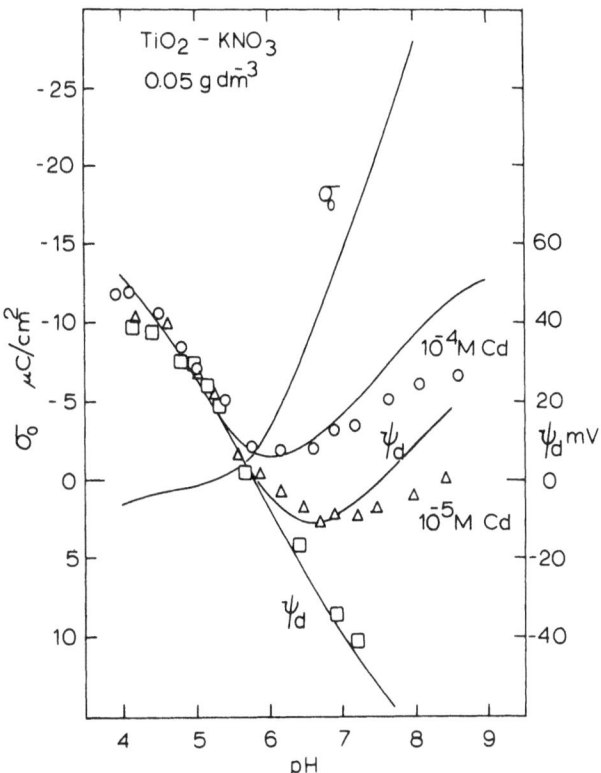

FIGURE 5   The zeta potential ($\zeta \equiv \psi_d$) of TiO$_2$ in 10^{-3} mol dm^{-3}
KNO$_3$ in the absence and presence of 10^{-5} and 10^{-4}
mol dm^{-3} Cd(NO$_3$) as a function of pH. The points are
experimental and the lines are calculated from the
model.   Also shown is the calculated surface charge
density but no data is available at such low TiO$_2$
concentrations.

$$SOH_2^+ \;\rightleftharpoons\; SOH + H_s^+; \quad K_{a_1}^{int} = \frac{[SOH][H_s^+]}{[SOH_2^+]} \tag{1}$$

and

$$SOH \;\rightleftharpoons\; SO^- + H_s^+; \quad K_{a_2}^{int} = \frac{[SO^-][H_s^+]}{[SOH]} \tag{2}$$

Thus, the density of hydroxyl sites (N_s sites/cm^2) dispersed in
solutions (S cm^2/dm^3) is equivalent to the analytical concentration of a reagent.

The intrinsic ionization constants, e.g., K_a^{int}, measure the tendency of the hydroxyl functional groups to undergo reactions. It has been argued that the strong dependence of the surface charge density on the background electrolyte concentration, C_{salt}, is in part due to ion pairing or surface complexation of the electrolyte ions with the surface hydroxyls, viz.,[23,25]

$$SOH_2^+NO_3^- \rightleftharpoons SOH + H_s^+ + NO_3^-{}_s, \quad {}^*K_{NO_3}^{int} = \frac{[SOH][NO_3^-][H_s^+]}{[SOH_2^+ \cdot NO_3^-]}$$

$$(3)$$

and

$$SOH + K_s^+ \rightleftharpoons SO^-K^+ + H_s^+, \quad {}^*K_K^{int} = \frac{[SO^-K^+][H_s^+]}{[SOH]K_s^+]} \qquad (4)$$

Again, the intrinsic complexation constants, e.g. ${}^*K_K^{int}$, measure the tendency for ion binding or exchange with the surface hydroxyls.

The ionization and reaction of sites leads to the formation of an electrical double layer at the interface. Ions are partitioned between the bulk solution and interface according to the Boltzman distribution, e.g.

$$[H_i^+] = [H^+] \exp\left(-\frac{e\psi_i}{kT}\right) \qquad (5)$$

where $[H_i^+]$ is the concentration of proton in the surface region i, $[H^+]$ is the bulk proton concentration and ψ_i is the electrostatic potential of the region. The adsorption of H^+ and OH^- defines the location of surface charge (σ_o) associated with the hydroxyl sites where the mean surface potential is ψ_o. The compact layer potential, ψ_b, is the mean potential in the plane imagined to contain adsorbed counter-ions, e.g. K_{aq}^+ at their closest distance of approach to the hydroxyl groups.[23,24]

Thus the concentrations or area density of surface sites may be expressed in terms of hydroxyl sites, solution concentrations, intrinsic constants and e.d.l. potential terms, e.g.

$$[SOH_2^+ \cdot NO_3^-] = [SOH][H^+][NO_3^-] \exp\left(\frac{e\psi_b - e\psi_o}{kT}\right) / {}^*K_{NO_3^-}^{int} \qquad (6)$$

$$[SOH_2^+] = [SOH][H^+] \exp\left(\frac{-e\psi_o}{kT}\right) / {}^*K_{a_1}^{int} \tag{7}$$

$$[SO^-] = K_{a_2}^{int}[SOH] \exp\left(\frac{e\psi_o}{kT}\right) / [H^+] \tag{8}$$

$$[SO^-K^+] = {}^*K_{K^+}^{int}[SOH][K^+] \exp\left(\frac{e\psi_o - e\psi_b}{kT}\right) / [H^+] \tag{9}$$

Since the protonic surface charge represents the net number of protons released or consumed by all surface reactions then

$$\sigma_o = \frac{F}{A}\left([SOH_2^+] + [SOH_2^+NO_3] - [SO^-] - [SO^-K^+]\right) \tag{10}$$

and the specifically adsorbed charge is

$$\sigma_b = \frac{F}{A}\left([SO^-K^+] - [SOH_2^+NO_3^-]\right). \tag{11}$$

Electroneutrality requires that

$$\sigma_o + \sigma_b + \sigma_d = 0 \tag{12}$$

where the diffuse layer charge density for flat plates in a symmetrical electrolyte is given by:

$$\sigma_d = -11.74 \times 10^{-6} \sqrt{C} \sinh \frac{e\psi_d}{2kT} \quad C/cm^2 \tag{13}$$

The potential and charge relationships in the compact double layer are given by

$$\psi_o = \psi_b + \frac{\sigma_o}{C_1} \tag{14}$$

and

$$\psi_b = \psi_d - \frac{\sigma_d}{C_2} \tag{15}$$

where C_1 and C_2 are the integral capacitances of the inner region of the e.d.l. This entire set of simultaneous equations for surface reactions and e.d.l. structure can be solved using the computer program MINEQL at any pH and analytical concentration with known values for the site density, ionization constants and e.d.l. capacitances C_1, C_2.

Table 1.   Intrinsic Ionization Constants and e.d.l. Data for TiO_2
            Hydrosols

$pK_{a_1}^{int}$	$pK_{a_2}^{int}$	$pK_{NO_{-3}}^{*int}$	$pK_{K^+}^{*int}$	N_S sites/nm^2	C_1 µF/cm^2	C_2 µF/cm^2
		A.	Batch CLD528			
2.7	9.1	4.5	7.2	12	110	20
		B.	Batch CLD640			
3.0	8.6	4.8	6.8	12	140	20

For the rutile sample, code named CLD528, Davis et al.[24] and James and Parks[1] used extrapolation techniques with Yates data[27,29] to estimate the intrinsic ionization and complexation constants and obtained agreement between model calculations and data. Their results are listed in Table 1. The number of exchangeable hydrogen ion sites N_S was determined by Yates[27,29] using tritium exchange.

The second rutile sample which was used for most of this work (batch number CLD640) had a slightly higher charge density. Calculations using slightly different intrinsic constants gave a reasonable simultaneous agreement with both the surface charge density and the zeta potential. The results of the calculations are shown in Figure 1 for both the surface charge density (σ_o) and the diffuse double layer potential ($\psi_d \sim \zeta$).

When cadmium nitrate is added in small amounts (2×10^{-4} mol dm^{-3} Cd) to the dispersion of TiO_2 (CLD528) in 10^{-2} mol dm^{-3} KNO_3, adsorption of cadmium occurs in the region of the pH_{pzc}, increasing to complete removal from the aqueous phase at high pH as indicated in Figure 2. The pH dependence alone demonstrates that OH^- is consumed or H^+ is released in the adsorption process. This is supported by the direct evidence from the potentiometric titration of the $TiO_2(CLD640)/KNO_3/Cd(NO_3)_2$ system in Figure 3. The types of reactions which could account for this behavior have already been suggested by several workers.[22,23] More specifically, Davis and Leckie[23] have reported a computer calculated isotherm for this particular system[24,26] in which the following adsorption or exchange reactions together with the reactions in Table 1a were assumed:

$$SOH + Cd^{2+}_{aq} \rightleftharpoons SO^-Cd^{2+} + H^+_{aq};$$

$$K^{int}_{Cd_1} = \frac{[SO^-Cd^{2+}][H^+]}{[SOH][Cd^{2+}]} \exp \frac{(2e\psi_b - e\psi_o)}{kT} \qquad (16)$$

and

$$SOH + Cd^{2+}_{aq} + H_2O \rightleftharpoons SO^-CdOH^+ + 2H^+_{aq};$$

$$K^{int}_{Cd_2} = \frac{[SO^-CdOH^+][H^+]^2}{[SOH][Cd^{2+}]} \exp \frac{(e\psi_b - e\psi_o)}{kT} \qquad (17)$$

Their criteria for selection of values for K^{int}_{Cd1} ($= 10^{-1.8}$) and K^{int}_{Cd2} ($= 10^{-8.7}$) was best fit to the Γ_{Cd} vs. pH adsorption isotherm using similar values for the other ionization and complexation reactions in Table 1a. In this analysis of our data, we have taken a wider approach in which we try to find those reactions and their characteristic constants that can be used to describe the whole range of direct adsorption, titrated surface charge and electrokinetic data.

We have found, principally by trial and error methods, that the most significant reaction that we can use is given by equation (17). In Figure 2 the solid line represents the result of calculations using the e.d.l. model described by equations (1) to (15), with parameter values in Table 1a (for TiO$_2$ CLD528) and with equation (17) using $pK^{int}_{Cd_2} = 9.3$.

This reaction scheme suggests that hydrolysis of the adsorbed cadmium may occur in the interfacial region in spite of the titration evidence in Figure 3, the reported ion-selective electrode response[26,28] and the literature values for hydrolysis constants which show that below pH 8 and 2×10^{-4} mol dm^{-3} Cd^{2+}, hydrolysis reactions such as

$$Cd^{2+} + H_2O \rightleftharpoons CdOH^+ + H^+; \qquad {}^*K_1 = 10^{-9.8} \qquad (18)$$

and

$$Cd(OH)_{2\ colloid} + 2H^+ \rightleftharpoons Cd^{2+} + 2H_2O; \qquad {}^*K_{so} = 10^{14.43} \qquad (19)$$

either are minor or do not occur.

The titration data in Figure 3 shows the dramatic consumption of OH⁻ associated with the adsorption of Cd onto TiO_2 (CLD640). At pH 8, the ratio of $OH_{consumed}/Cd_{adsorbed}$ is 2. It should be noted that this stoichiometry is given by reaction scheme (17) where two protons are released for each adsorbed cadmium species.

From Figure 3, we have obtained the equivalent charge density of protons released or consumed by surface reactions. This is plotted as the square symbols in Figure 4. In the computational model used in MINEQL, the surface charge density is evaluated by the proton balance for hydroxylated sites [SOH], so that our notation

$$\sigma_o = F\left(\Gamma_{SOH_2^+NO_3^-} + \Gamma_{SOH_2^+} - \Gamma_{SO^-} - \Gamma_{SO^-K^+} - \Gamma_{SO^-CdOH^+} \right) C/cm^2$$

(20)

The circle symbols in Figure 4 represent the values of σ_o derived from the σ_H data for the TiO_2 (CLD640)/KNO_3/Cd$(NO_3)_2$ system. Also shown is a solid line which represents the results of the model calculations using equations (1) to (15), Table 1b for TiO_2 (CLD640) and equation (17) with the value $pK_{Cd_2}^{int}$ = 9.3. We note the good agreement obtained between the experimentally derived value and the calculated result for σ_o. Also shown, but without any experimental data, is the calculated diffuse double layer potential ψ_d for these conditions of high TiO_2 powder concentration. Mass transport electrophoresis would be required to experimentally investigate such a system at relatively high particle concentration.

The electrophoretic mobility of dilute TiO_2 (CLD640) suspensions was measured in several cadmium nitrate, 10^{-3} mol dm⁻³ KNO_3 solutions. The mobility data was converted to zeta potential[24] using the Wiersema, Loeb and Overbeek method[34] and the resulting zeta potential of TiO_2 is shown in Figure 5, as a function of pH for no added Cd$(NO_3)_2$ and for added concentration of 10^{-5} and 10^{-4} mol dm⁻³. The experimental results are compared to lines which represent the calculated diffuse layer potential, ψ_d described in the model together with equations (1) to (15), characteristic constants in Table 1b (TiO_2, CLD640) and equations (17) with $pK_{Cd_2}^{int}$ = 9.3. On inspection of equation (17) in which it is proposed that the site formed could be written as $SO^- \cdot CdOH^+$, it would appear difficult to explain the reversal of the zeta potential in the presence of cadmium ions. However, there are at least five simultaneous reactions occurring and the formation of sites such as $SO^- \cdot CdOH^+$ affects the equilibrium position of the others, and in this model this causes an excess of SOH_2^+ sites over SO^- sites. This excess charge is not balanced in the inner part of the e.d.l. but by the diffuse layer charge, σ_d, and of course is reflected in a reversal of the zeta potential or ψ_d.

Table 2.   Summary of Reactions and Experiment Conditions for
            Description of the Electrokinetic, Charge and Adsorption
            Properties of TiO$_2$ Hydrosols

$$TiOH_2^+ \rightleftharpoons TiOH + H_s^+ \qquad\qquad pK_{a_1}^{int} = 3.0$$

$$TiOH_2^+NO_3^- \rightleftharpoons TiOH + H_s^+ + NO_{3_s}^- \qquad pK_{NO_3}^{*int} = 4.8$$

$$TiOH \rightleftharpoons TiO^- + H_s^+ \qquad\qquad pK_{a_2}^{int} = 8.6$$

$$TiOH + K^+ \rightleftharpoons TiO^-K^+ + H_s^+ \qquad pK_{K^+}^{*int} = 6.8$$

$$TiOH + Cd^{2+} + H_2O \rightleftharpoons TiO^-\cdot CdO^+H + 2H^+ \qquad pK_{Cd}^{*int} = 9.3$$

$$C_1 = 140 \ \mu F/cm^2, \qquad C_2 = 20 \ \mu F/cm^2$$

$$12.0 \ sites/nm^2 \ (N_H)$$

Method	TiO$_2$	C_{KNO_3} (mol dm^{-3})	$C_{Cd(NO_3)_2}$ (mol dm^{-3})
ζ	0.05 g/dm^3	10^{-3}	$-, 10^{-5}, 10^{-4}$
σ_H, σ_o	5.0 g/dm^3	$10^{-3}, 10^{-2}, 10^{-1}$	$-, 10^{-4}$
Γ_{Cd}	10.0 g/dm^3	10^{-2}	2×10^{-4}

    As in all the comparisons between the various experimental
quantities and their calculated values, the agreement is good but
not perfect. Taken individually each comparison might be regarded
as interesting but not outstanding. However, taken as a whole,
i.e. a simulation of one of the few complete sets of adsorption
isotherms, surface charge determination and electrokinetic potentials in the literature, we feel there is a strong argument for
the induced surface hydrolysis reactions of the type

$$SOH + Cd_s^{2+} + H_2O \rightleftharpoons SO^-CdOH^+ + 2H_s^+ . \qquad (21)$$

Other reaction schemes may be used to fit adsorption data or
other properties alone but within the limits imposed by our trial
and error search, reaction (17) appears to be of unique importance.
With this reaction scheme the properties Γ_{Cd}, σ_o (or σ_H) and
$\psi_d(\zeta)$ have been simulated over the concentration ranges from 0.05
to 10 g dm^{-3} TiO$_2$, 10^{-5} to 2 x 10^{-4} mol dm^{-3} Cd(NO$_3$)$_2$ and 10^{-3}
to 10^{-1} mol dm^{-3} KNO$_3$, background electrolyte, in the pH interval
from 4 to 9. The reactions considered and the range of conditions
such as salt concentration TiO$_2$ concentration and Cd(NO$_3$)$_2$
concentration for the various experiments are given in Table 2.
Of course, as Davis and Leckie have already pointed out,[25]
reaction (16) improves the agreement for adsorption isotherms,
Γ_{Cd} vs. pH, in the lower pH range. It would seem that a whole
range of surface induced hydrolysis reactions could be involved
with some being more dominant than others. Certainly, if the pH
of aqueous metal ion-oxide dispersions is increased sufficiently
by addition of more base, in this particular case up to and above
pH 9, the electrophoretic mobility results indicate the formation
of a coating of the hydrous oxide of the adsorbed metal ion,
e.g. Cd(OH)$_2$.[26] Such a hydrous oxide coating has its own ionization characteristics, pzc and iep. Whether this coating occurs
through condensation of adsorbed hydrolysed species, or by heterocoagulation between oxide substrate and freshly precipitated
hydrous oxides, e.g. TiO$_2$ and Cd(OH)$_2$, will depend on the way the
system was prepared and altered by addition of reagents, e.g.
acids or bases.

Our studies have been complicated by the use of two batches
of TiO$_2$, which appear to have slightly different ionization
characteristics. It is to be hoped that future studies will be
planned in which model colloids, e.g. spherical mono-dispersed[10,35]
TiO$_2$ colloid particles, can be produced in sufficient quantities
to allow the complete range of experiments for characterization
of the surface reactions of aqueous colloids. Also other complementary techniques such as XPS and IR spectroscopy are needed to
help identify surface coatings where these measurements can be
made in situ without disturbing the system or at minimum perturba-
tion of the aqueous hydrosol.

B.   Effect of Parameter Values on the Behavior of Model Colloids

Throughout this discussion, we have stressed the importance
of using a variety of techniques to investigate aqueous colloids
because of the difficulties involved in characterizing the adsorbed
species. Models such as the one described here can be very
useful in testing hypotheses about the form of the adsorbed ions.
However, interpretations based on such an approach will always be
model dependent. Thus, it is useful to know about the sensitivity
of the calculated results to variations in values of parameters

Table 3.   Hypothetical Intrinsic Constants for Model
           Colloids

Model Colloid	$pK_{a_1}^{int}$ $pK_{a_2}^{int}$	ΔpK_a	pK_{Cl}^{*int} pK_{Na}^{*int}	ΔpK_X
A	5.5 8.5	3	5.5 8.5	3
B	4.0 10.0	6	5.5 8.5	3
C	3.0 11.0	8	5.5 8.5	3

$$N_S = 10 \text{ sites/nm}^2$$

$$C_1 = 100 \text{ } \mu F/cm^2 \qquad C_2 = 20 \text{ } \mu F/cm^2$$

$$pH_{pzc} = pH_{iep} = 7.0$$

used in the model, e.g. intrinsic ionization or binding constants.
To illustrate this sensitivity to ionization constants, we shall
consider calculations based on an ideal model colloid with selected
values of the intrinsic constants.

   These model colloids might have ionization and complexation
properties such as are given in Table 3. Calculations for the
surface charge density σ_0, based on the model described by equations (1) to (15) with the values given in Table 2, were performed
and are plotted in Figure 6 as a function of pH at a number of
electrolyte concentrations. It is immediately noticed that the
surface charge density (σ_0) is not very sensitive to the ionization constants pK_{a1}^{int} and pK_{a2}^{int}, since all the σ_0-pH results are
essentially the same for quite a wide range of values for pK_{a1}^{int}
and pK_{a2}^{int}. In addition to the calculation of σ_0, we simultaneously
obtained the diffuse layer potential, ψ_d (equations (13) and
(15)). The results for ψ_d are plotted as a function of pH and
electrolyte concentrations in Figures 7 and 8 for model colloids
B and C of Table 3. Here it is seen that while σ_0 is not very

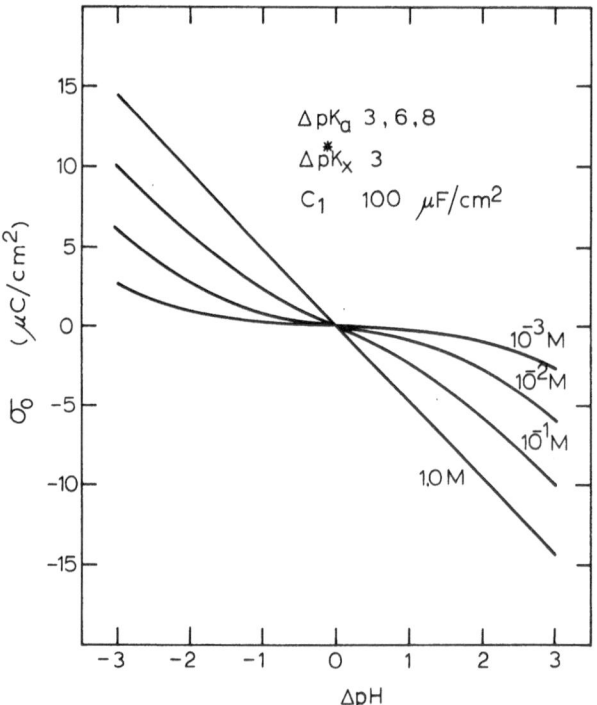

FIGURE 6   Calculations of the surface charge density of model
amphoteric colloids, A, B and C predicted by the use of
the e.d.l. model at various ionic strengths as a
function of pH for a number of values of ionisation
constants ΔpK_a = 3,6,8.   (See Table 3).

sensitive to pK_a^{int} values, the diffuse layer potential ψ_d is
quite sensitive to pK_a^{int} values with the magnitude of ψ_d becoming
smaller as the difference in pK_a values (= $\Delta_p K_a$) becomes greater.
This shows that if we are to rely on computer based models for
interpretation of surface reactions, then the analysis should be
based on both σ_o and $\psi_d(\zeta)$ measurements.   The reason for the
decreasing magnitude of ψ_d as the separation of pK_a^{int} values
increases is the implied accompanying increase in the binding of
counter-ions since pK_{Na} binding = pK_{a2}^{int} - pK_{Na}^{int}.   Thus for model
colloid A, pK_{Na} binding is zero, for colloid B, it is -1.5 and
for colloid C, it is -2.5.   As a consequence of stronger binding
more counter-ion charge is located in the inner compact layer;
hence σ_b increases closer in magnitude to σ_o.   Consequently the
charge on the diffuse layer (σ_d) and its potential (ψ_d) decrease
in magnitude.   In comparison to observed zeta potential behavior,
the magnitude and shape of the ζ - pH dependence for model C in
Figure 8 are not common and this suggests that the binding tendency
of counter-ions is not as great as implied by the selected value

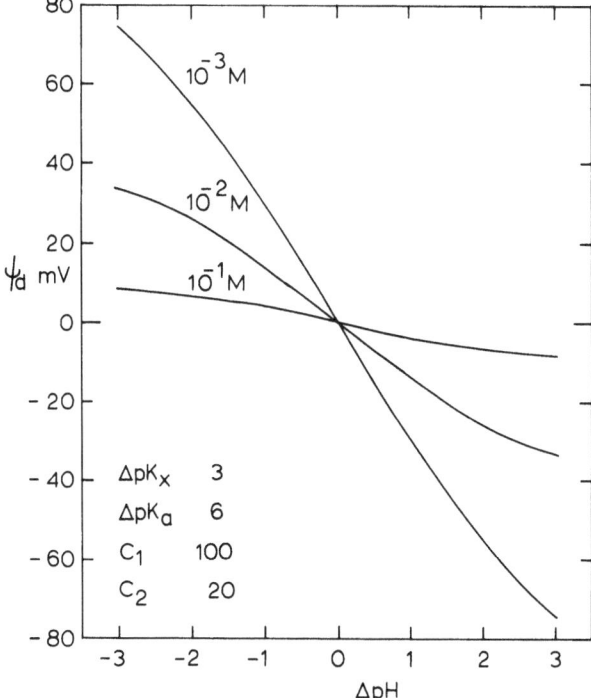

FIGURE 7   Calculations of the zeta potential ($\zeta \equiv \psi_d$) of a model
amphoteric colloid   B ($\Delta pK_a = 6$) as a function of pH at
a number of ionic strengths.

of colloid C in Table 3.   However, the calculated values of σ_o
and $\psi_d(\zeta)$ for colloid B are much more in line with experimental
data.

In addition to the intrinsic ionization constants, this type
of computer model requires values to be assigned to the interfacial capacitances C_1 and C_2. C_2 has been kept constant at
20 $\mu F/cm^2$ which is the magnitude of capacitance at Hg/H₂O and
AgI/H₂O interfaces.   On the other hand C_1 has been varied to
improve the agreement of calculated values of σ_o and ζ with
experimental data.   Values typically used are around 100 - 140
$\mu F/cm^2$ which is about twice that expected from Hg/H₂O and AgI/H₂O
interfaces.   The effect of increasing C_1 is to increase the slope
of the σ_o - pH curves for each ionic strength but with the largest
effect at higher ionic strength.   Experimentally σ_o - pH curves
tend to become linear at the higher ionic strengths e.g. > 1 M.
This enables estimation of the minimum value of the inner layer
capacitance from the slope, since $C = d\sigma_o/d\psi_o = d\sigma_o/dpH$. $dpH/d\psi_o$
and the maximum value of $d\psi_o/dpH$ is 59.2 mV at 25°C.

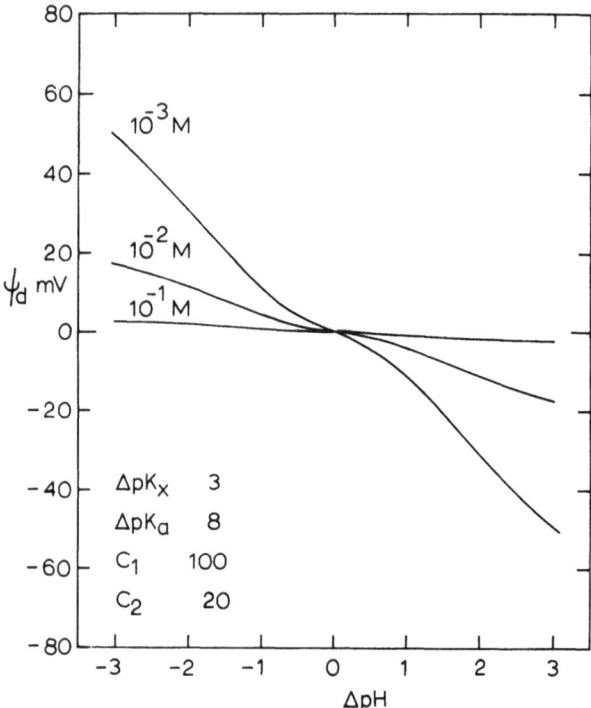

FIGURE 8   Calculations of the zeta potential ($\zeta \equiv \psi_d$) of a model
amphoteric colloid   C (ΔpK_a = 8) as a function of pH at
a number of ionic strengths.

In connection with experimental evaluation of the interfacial
capacitance, Tomkiewicz[36] has used a doped TiO_2 single crystal
electrode to evaluate the capacitance of the Hemholtz layer, C_H,
as a function of pH in various mixed electrolyte solutions.   For
example, in 0.1 M CH_3COOH/Na_2SO_4 electrolyte at pH 2.7, C_H =
64 ± 20 $\mu F/cm^2$.   This reflects a combination of inner and outer
compact layer capacitances C_1 and C_2.

There appears to be opportunity for future work in which the
ionization constants and capacitances are optimised using experimental surface charge, adsorption density and zeta potential
data.

REFERENCES

1.   R.O. James and G.A. Parks, Characterization of aqueous
     colloids by their electrical double layer and intrinsic
     surface charge properties, in "Surface and Colloid Science",
     Vol. 12, in press, Plenum Press, 1981.

2. E.A. Jenne, Chemical modeling - goals, problems, approaches and priorities, in "Chemical Modeling in Aqueous Systems - Speciation, Sorption, Solubility and Kinetics", Ed. E.A. Jenne, A.C.S. Symposium series No. 93, Washington, DC, 1979.

3. J.W. Murray and J.G. Dillard, Geochemica et Cosmochemica Acta 43:781 (1979).

4. M.H. Koppelman and J.W. Dillard, J. Colloid Interface Sci. 66:345 (1978).

5. P.H. Tewari and W. Lee, J. Colloid Interface Sci. 52:77 (1975).

6. P.H. Tewari and N.S. McIntyre, "Characterization of Adsorbed Cobalt at the Oxide/Water Interface", AIChE Symposium Series 71:150 (1975).

7. M. Ashida, M. Sasaki, H. Kan, T. Yasunaga, K. Hachiya and T. Inoue, J. Colloid Interface Sci. 67:219 (1978).

8. K. Hachiya, M. Ashida, M. Sasaki, H. Kan, T. Inoue and T. Yasunaga, J. Phys. Chem. 83:1866 (1979).

9. G. Scatchard, New York Acad. Sciences 57:660 (1949).

10. M. Visca and E. Matijevic, J. Colloid Interface Sci. 68:308 (1979).

11. E. Matijevic and P. Scheiner, J. Colloid Interface Sci. 63:509 (1978).

12. E. Matijevic, R.S. Sapieszko and J.B. Melville, J. Colloid Interface Sci. 50:567 (1975).

13. R. Demchak and E. Matijevic, J. Colloid Interface Sci. 31:257 (1969).

14. W.B. Scott and E. Matijevic, J. Coll. Interface Sci. 66:447 (1978).

15. R. Brace and E. Matijevic, J. Inorg. Nucl. Chem. 35:3691 (1973).

16. J.W. Goodwin, J. Hearn, C.C. Ho and R.H. Ottewill, Colloid & Polymer Sci. 252:464 (1974).

17. Y. Chung-Li, J.W. Goodwin and R.H. Ottewill, Prog. Colloid & Polymer Sci. 60:163 (1976).

18. J.W. Goodwin, R.H. Ottewill and R. Pelton, Colloid Polym. Sci. 257:61 (1979).

19. P. Bagchi, B.V. Gray and S.M. Birnbaum, J. Colloid Interface Sci. 69:502 (1979).

20. A. Homola and R.O. James, J. Colloid Interface Sci. 59:123 (1977).

21. R.M. Fitch and W.T. McCarvill, J. Colloid Interface Sci. 66:20 (1978).

22. W.T. McCarvill and R.M. Fitch, J. Colloid Interface Sci. 64:403 (1978).

23. R.O. James, J.A. Davis and J.O. Leckie, J. Colloid Interface Sci. 65:331 (1978).

24. J.A. Davis, R.O. James and J.O. Leckie, J. Colloid Interface Sci. 63:480 (1978).

25. J.A. Davis and J.O. Leckie, J. Colloid Interface Sci. 67:90 (1978).

26. P.J. Stigich, "Adsorption of Cadmium (II) Complexes at the Oxide/Water Interface", M.Sc. Thesis University of Melbourne, 1976.
27. D.E. Yates, "The Structure of the Oxide/Aqueous Electrolyte Interface", Ph.D. Thesis University of Melbourne, 1975.
28. R.O. James, P.J. Stiglich and T.W. Healy, Faraday Discussions of the Chemical Society 59:142 (1975).
29. D.E. Yates, R.O. James and T.W. Healy, J.C.S. Faraday Trans. I, 76:1,9 (1980).
30. G.R. Wiese and T.W. Healy, J. Colloid Interface Sci. 51:421 (1975).
31. G.R. Wiese, "Cation Adsorption and Heterocoagulation in Oxide-Water Systems", Ph.D. Thesis University of Melbourne, 1973.
32. J. Drzymala, J. Lekki and J. Laskowski, Colloid & Polymer Sci. 257:768 (1979).
33. P.W. Schindler, E. Walti and B. Furst, Chimia 30:107 (1976).
34. P.H. Wiersema, A.L. Loeb and J.Th.G. Overbeek, J. Colloid Interface Sci. 22:78 (1966).
35. E. Matijevic, M. Budnik and L. Meites, J. Colloid Interface Sci. 61:302 (1977).
36. M. Tomkiewicz, J. Electrochem. Soc. 126:1505 (1979).

EFFECTS OF STRONG BINDING OF ANIONIC ADSORBATES ON ADSORPTION

OF TRACE METALS ON AMORPHOUS IRON OXYHYDROXIDE

Mark M. Benjamin and Nickolas S. Bloom

Department of Civil Engineering
University of Washington
Seattle, Washington 98195

ABSTRACT

Adsorption of three metal ions (Cd^{2+}, Co^{2+}, Zn^{2+}) onto amorphous iron oxyhydroxide has been studied in the presence of several strongly binding anionic adsorbates (SeO_4^{2-}, SeO_3^{2-}, AsO_4^{3-}, AsO_3^{3-}, CrO_4^{2-}, PO_4^{3-}). The anions either increase or have no effect on trace metal adsorption, indicating that competition between the anions and cations for surface sites is relatively minor. The results cannot be explained by electrostatic interactions unless the relative acidities of the anions are very different at the surface than in solution. The most likely cause of the enhanced metal adsorption in some systems is that a secondary surface phase forms, and the metals bind more strongly to the new phase than they do to the original substrate. The secondary phase appears to be an iron-anion solid rather than a trace metal-anion solid. In systems where a new surface does not form, electrostatic and competitive interactions are somehow suppressed, which may indicate that anion adsorption sites are physically and electrically isolated from cation adsorption sites.

INTRODUCTION

Adsorption onto particulate matter influences the soluble concentration, transport rate, and bioavailability of numerous chemical species including heavy metals, pesticides, some nutrients, natural and anthropogenic organic compounds, and many radionuclides.[1-3] Among the most intensively studied and important adsorption reactions are those of heavy metals and oxyanions on metal oxide surfaces.[4,5] Environmental contaminants such as

Cu, Cd, Pb, and As are mobilized in industrial processes or by
natural weathering and eventually may be discharged to an aquatic
system. Adsorption reactions occurring in waste treatment processes,
rivers, estuaries or groundwater systems may control the fate of
these potential toxins from that time on.

Oxide surfaces in aqueous suspension act as weak diprotic
acids, with each oxide surface group ($\equiv$ SO) capable of binding
zero, one or two protons in response to the activity of protons
and other potential adsorbates in solution:[6]

$$\equiv SO^- \xrightleftharpoons[-H^+]{+H^+} \equiv SOH^0 \xrightleftharpoons[-H^+]{+H^+} SOH_2^+ \qquad (1)$$

Oxide groups may also bind metal ions, anions, or soluble metal-
ligand complexes, and in some cases a single adsorbate molecule
may bind to two or more oxide sites (multi-dentate binding).
Thus, adsorption reactions are similar to acid-base and complexation reactions in solution, and may be described by surface
acidity or surface complexation stability constants. The main
difference between solution and surface reactions is the non-zero
electrical potential in the interface region, and this potential
must be taken into account when evaluating the surface binding
constants.[7,8]

Some workers have calculated the various energy terms contributing to the overall adsorption energy, and based on these calculations have tried to infer the speciation of adsorbed molecules.[9,10] However, the calculations and conclusions always
depend on assumptions made about the location of adsorbed ions
and the charge-potential relationship in the interphase. The
question of whether adsorbed species are hydrolyzed or unhydrolyzed, partially or completely hydrated, and mono- or bi-dentate,
are therefore still unresolved. Nevertheless, several self-
consistent mathematical models have been developed which reproduce
experimental results for adsorption of trace metals and oxyanions
extremely well.[11-13] Some of these models can also describe
adsorption of bulk electrolyte ions and surface electrical properties. For the most part, these models have been developed and
tested based on data for systems in which only one ion is strongly
adsorbed, i.e., adsorbed to such an extent that its concentration
in solution is significantly reduced or until the available
surface sites are almost all filled. In a few cases competitive
adsorption has been studied involving two similar, strongly
binding ions, e.g., Cd^{2+} and Cu^{2+}; SeO_4^{2-} and SO_4^{2-}.[14,15] The
models have reproduced experimental results accurately in some of
these systems but not in others. The present work was undertaken

to investigate surface interactions among dissimilar ions, specifically those between strongly binding anionic and cationic
adsorbates.

METHODS AND MATERIALS

All adsorption experiments were conducted in 0.1 M $NaNO_3$
solutions using amorphous iron oxyhydroxide ($Fe_2O_3 \cdot H_2O$(am)) as
the adsorbent. This solid was precipitated in situ and aged two
hours at pH 7 before any adsorbates were added to the system.
Adsorbates were then added along with radioactive tracers, pH was
adjusted, and the slurries were equilibrated two hours. Fractional adsorption was determined by crystal scintillation counting
of the supernatant from the centrifuged slurries. All other
procedures were identical to those used by Benjamin and Leckie.[16]
The surface properties of this solid have been reported by Davis
and Leckie.[13]

RESULTS AND DISCUSSION

Possible Surface Interactions Between Competing Adsorbates

When more than one strongly binding adsorbate is present in
a system, the adsorption of each ion may affect the others in
several ways. Since part of the driving force for adsorption is
electrostatic interaction between the adsorbate and the surface,
binding of any ion to the surface will alter the tendency of
another ion to bind. Let A be the ions whose adsorption behavior
is being investigated, and B be the "competing" ion which is present
in the system. Adsorption of B may enhance or diminish the
electrostatic contribution for adsorption of A. The direction
and magnitude of this effect depend on the relative charges on A
and B, and the complete stoichiometry of the reactions. The net
charge of A or B in solution may suggest how each will affect the
surface potential when it adsorbs. However, it is important to
recognize that the charge on the dominant adsorbed ion may not be
the same as that on the dominant dissolved ion. For instance,
shown below are three potential adsorption stoichiometries for
adsorption of B^{2+} ions from solution which increase, do not
affect, and decrease charge in the interfacial region.

Change in Interface
<u>Electrical Potential</u>

$$\equiv SO^- - A^+ + B^{2+}$$

$$\longrightarrow \; \equiv SO^- - B^{2+} + A^+ \qquad\qquad \text{increase} \qquad (2)$$

$$\equiv SO^- - A^+ + B^{2+} + H_2O$$

$$\longrightarrow \; \equiv SO^- - BOH^+ + A^+ + H^+ \qquad\qquad \text{no effect} \qquad (3)$$

$$\equiv SO^- - A^+ + B^{2+} + H_2O$$

$$\longrightarrow \; \equiv SO^- \!\!-\!\! BOH^+ + A^+ + H^+ \qquad\qquad \text{decrease} \qquad (4)$$

In Reaction (2), surface potential increases because a divalent ion replaces a monovalent ion. In Reaction (3), the divalent ion hydrolyzes when it adsorbs, so there is no net change in surface charge or potential. However the solution pH will change. The stoichiometry of Reaction (4) is the same as that of Reaction (3), but the adsorbed B ions are farther from the surface than in Reaction (3). This causes a slight, but non-zero decrease in electrical potential in the interphase region in Reaction (4). Numerous other reactions could be proposed to demonstrate the same point: the charge on an ion in solution is not necessarily a good indicator of how it will affect surface potential when it adsorbs. This is an extremely important point in competitive adsorption studies, and means that it is at least theoretically possible for sorption of one cation to enhance sorption of another cation or to diminish sorption of an anion.

In addition to electrical interactions, strongly binding adsorbates may compete for available binding sites. This type of interaction always reduces the tendency for each ion to adsorb. If the system obeys the Langmuir isotherm, adsorption of A in the non-competitive and competitive systems is given by:

$$\Gamma_A = \frac{\Gamma_{M,A}\, K_A\, C_A}{1 + K_A C_A + \displaystyle\sum_{i=B} \eta_i K_i C_i}$$

where Γ_A = adsorption density of A

$\Gamma_{M,i}$ = maximum adsorption density of component i

K_i = Langmuir binding constant for component i

C_i = equilibrium soluble concentration of component i

η_i = $\Gamma_{M,A}/\Gamma_{M,i}$

The summation includes all competing adsorbates in the system. The factor η_i corrects for the fact that some adsorbates may take up more space or more sites than others. It is rarely considered in the classical derivation of competitive adsorption isotherms, since most studies of competitive adsorption have involved species of similar size ($\eta = 1.0$). However, it may be important when species of widely varying size compete for sites.

A third type of ion-ion interaction in these systems is bridging. It is possible that specific chemical interactions will allow ion B to bind simultaneously to the surface and to ion A, thereby removing A from solution even though no surface-to-A bond is formed. This situation can be described by two limiting cases. In one case, a soluble A-B complex binds as a single entity or is formed at the surface, yielding a discrete adsorbed species. In the other case, ion B may form a surface precipitate to which ion A adsorbs. In this case the presence of the original adsorbent becomes irrelevant, except as a site for precipitation of B. Adsorption of A is then expected to be similar to its adsorption in a system containing only precipitated B and no other solid.

Adsorption in Non-Competitive Systems

Adsorption of individual cations (Cd^{2+}, Co^{2+}, Zn^{2+}) and anions (SeO_4^{2-}, SeO_3^{2-}, CrO_4^{2-}) on $Fe_2O_3 \cdot H_2O$(am) was investigated for several different concentrations of adsorbate. Typical pH-adsorption curves for cations and anions in systems with only one strongly binding adsorbate (non-competitive systems) are presented in Figures 1-3. For most metal ions, fractional adsorption (amount of metal adsorbed/total metal in the system) decreases with increasing total metal concentration in the system even when surface sites are available in excess. This phenomenon has been interpreted as indicating the presence of multiple types of surface sites.[12] Anionic adsorbates exhibit a similar trend, though it is somewhat less pronounced. The results for all non-competitive adsorption systems agree well with those reported by Leckie and co-workers.[12-16]

These results have been modeled using the Stanford Generalized Model for Adsorption (SGMA) developed by Davis et al.[7] In addition, we have modeled adsorption of AsO_4^{3-} on $Fe_2O_3 \cdot H_2O$(am) based on

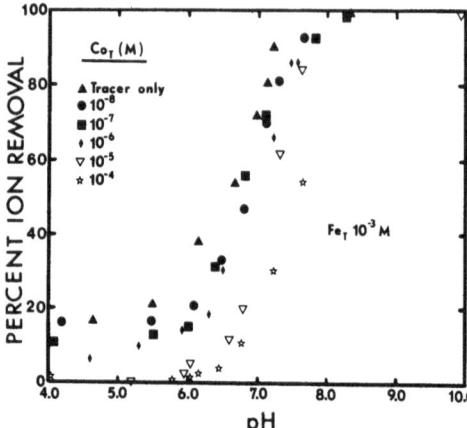

FIGURE 1: Adsorption of Co on $Fe_2O_3 \cdot H_2O$(am) as a function of pH
         for a wide range of total Co concentrations. The trend
         of the adsorption edge shifting to more alkaline pH
         with increasing total metal was observed for Cd and Zn
         as well.

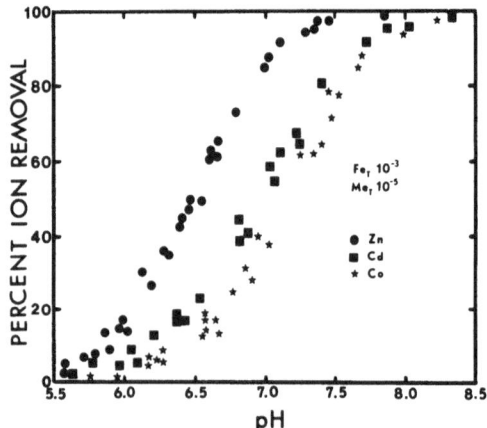

FIGURE 2: Adsorption of Co, Zn, and Cd on $Fe_2O_3 \cdot H_2O$(am) as a
         function of pH for 10^{-5} M total metal in the system.

literature data.[16] The modeling involves choosing one or two
reasonable adsorption stoichiometries and adjusting the stability
constants for these reactions until the model results match the
experimental results. If a good match cannot be obtained, a new
stoichiometry is chosen and another attempt is made. The modeling
results are summarized in Table 1. For both cations and anions,
the best fit results when the stability constant for the dominant
adsorption reaction varies as a function of adsorption density.

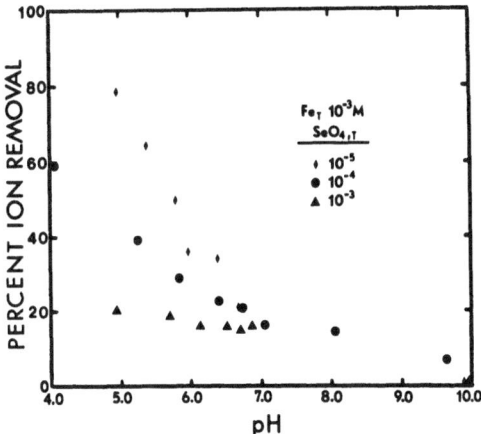

FIGURE 3: Adsorption of SeO_4^{2-} as a function of pH and total
          SeO_4^{2-} added to solution.

Such a variation is inconsistent with theory. Possible explanations for the variation are that chemical differences exist among
groups of surface sites, that some process other than adsorption
is controlling partitioning of the ions in these systems, or that
there is a flaw in the formulation of the model. A detailed
discussion of this problem has been presented elsewhere.[12] The
model also indicates that each anion must bind or block access to
more than one surface site, which is reasonable based on the size
of these oxyanions.[14]

Adsorption in Competitive Systems

Results of adsorption of 10^{-5} M Cd, Co, or Zn on $Fe_2O_3 \cdot H_2O$(am)
in the presence of several different oxyanions are presented in
Figures 4 through 9. For each metal the presence of the anions
either has no effect or enhances adsorption. In all cases, the
magnitude of the effect increases in the order $0 \simeq SeO_4^{2-} < AsO_3^{3-}$
$\leq CrO_4^{2-} < SeO_3^{2-} < AsO_4^{3-} \leq PO_4^{3-}$. The results are surprising in
a number of ways. First, the adsorption behavior of SeO_4^{2-},
CrO_4^{2-}, and SeO_3^{2-} in non-competitive systems indicates that in
some experiments (10^{-3} M total anion, 10^{-3} M Fe) a large fraction
($> 60\%$) of the surface sites is occupied or blocked by the anions.
This conclusion is based on our data and that of Davis and
Leckie,[14] and Davis and Leckie's estimates that 3 to 4 surface
sites are blocked each time one anion molecule adsorbs. For
instance, Figure 3 indicates that at pH 6.0, $\sim$ 17 percent of the
SeO_4^{2-} is adsorbed, in a system containing 10^{-3} M Fe and 10^{-3} M
total SeO_4^{2-}. Estimating surface site density on $Fe_2O_3 \cdot H_2O$(am) to

Table 1. Best fit intrinsic surface complexation constants for adsorption on $Fe_2O_3 \cdot H_2O$(am) based on the SGMA.

Total Adsorbate Added	Reaction[a]	Stability Constant[b]
10^{-5} Cd^{2+}	I	$10^{-4.80}$
	II	$10^{-11.25}$
10^{-5} Co^{2+}	I	$10^{-4.80}$
	II	$10^{-11.60}$
10^{-5} Zn^{2+}	I	$10^{-4.80}$
	II	$10^{-10.50}$
10^{-4} SeO_4^{2-}	III	$10^{11.75}$
	IV	$10^{15.60}$
10^{-3} SeO_4^{2-}	III	$10^{15.00}$
	IV	$10^{19.00}$
10^{-4} SeO_3^{2-}	III	$10^{12.80}$
	IV	$10^{20.75}$
10^{-3} SeO_3^{2-}	III	$10^{12.90}$
	IV	$10^{22.00}$
10^{-5} CrO_4^{2-}	III	$10^{11.90}$
	IV	$10^{18.00}$
10^{-4} CrO_4^{2-}	III	$10^{14.40}$
	IV	$10^{16.80}$
10^{-3} CrO_4^{2-} [c]	-	-
	-	-
10^{-4} AsO_4^{3-} [d]	V	$10^{27.70}$
	VI	$10^{33.50}$
10^{-3} AsO_4^{3-} [d]	V	$10^{27.70}$
	VI	$10^{33.50}$

Table 1, concluded

[a]Stoichiometries for the model reactions are as follows: The notation corresponds to that used by Davis et al.[7] in their development
of the SGMA.

$$\text{I} \quad SOH^0 + Me^{2+} \longrightarrow SO^- - Me^{2+} + H^+$$

$$\text{II} \quad SOH^0 + Me^{2+} + H_2O \longrightarrow SO^- - MeOH^+ + 2H^+$$

$$\text{III} \quad SOH^0 + A^{2-} + H^+ \longrightarrow SOH_2^+ - A^{2-}$$

$$\text{IV} \quad SOH^0 + A^{2-} + 2H^+ \longrightarrow SOH_2^+ - HA^-$$

$$\text{V} \quad SOH^0 + B^{3-} + 2H^+ \longrightarrow SOH_2^+ - HB^{2-}$$

$$\text{VI} \quad SOH^0 + B^{3-} + 3H^+ \longrightarrow SOH_2^+ - H_2B^-$$

[b]For systems containing 10^{-3} M Fe_T, and assuming one site occupied
per cation and three sites occupied per anion adsorbed, unless
otherwise noted.

[c]The data for this system cannot be modeled unless fewer than three
sites are occupied per CrO_4^{2-} ion adsorbed. The assumption of three
sites per ion leads to an apparent site occupancy > 100%.

[d]The AsO_4^{3-} constants are based on the data in Reference (16), from
a system with 5×10^{-5} M $AsO_{4,T}^{3-}$ and 4×10^{-4} Fe_T.

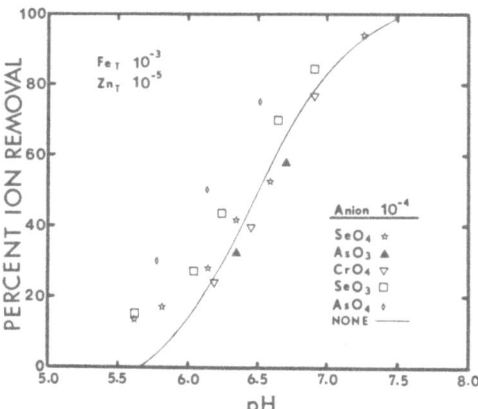

FIGURE 4: Effect of various anions on Zn adsorption for 10^{-4} M anion added.

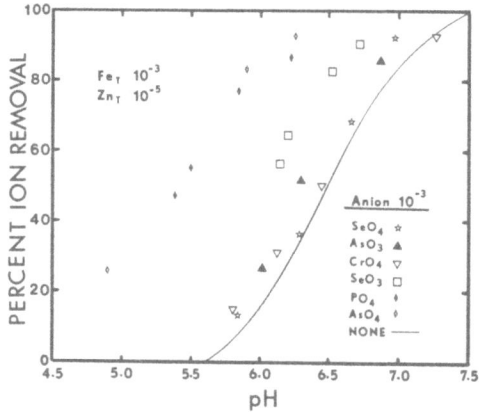

FIGURE 5: Effect of various anions on Zn adsorption for 10^{-3} M anion added.

be 0.87 sites per Fe atom,[13] the fraction of sites occupied under these conditions is:

% occupied

$$= \frac{(17\% \text{ ads.})(10^{-3} \text{ M SeO}_{4,T}^{2-})(3 \text{ sites occupied/SeO}_{4}^{2-} \text{ ads.})}{(10^{-3} \text{ M Fe}_{T})(0.87 \text{ sites/Fe atom})} = 59\%$$

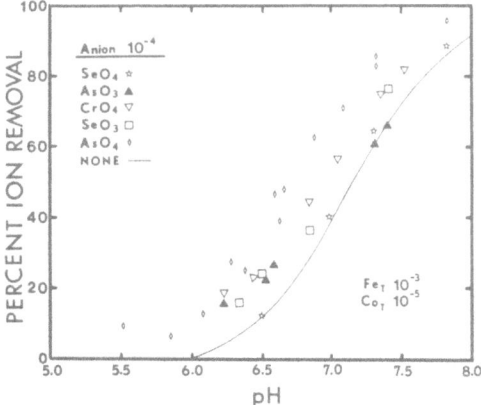

FIGURE 6: Effect of various anions on Co adsorption for 10^{-4} M anion added.

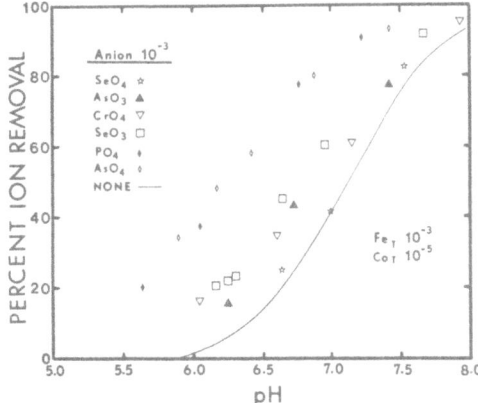

FIGURE 7: Effect of various anions on Co adsorption for 10^{-3} M anion added.

If the surface is indeed occupied to this extent by the anions, adsorption of the cations should be significantly reduced due to competition for sites. Yet in no case did addition of anions to the system decrease cation adsorption at all.

Of course, it is possible that adsorption of anions enhances electrostatic interaction between the surface and the metals so much that competitive interactions are overwhelmed. However, several considerations argue against this explanation. First, in several systems, the anions did not affect trace metal partitioning

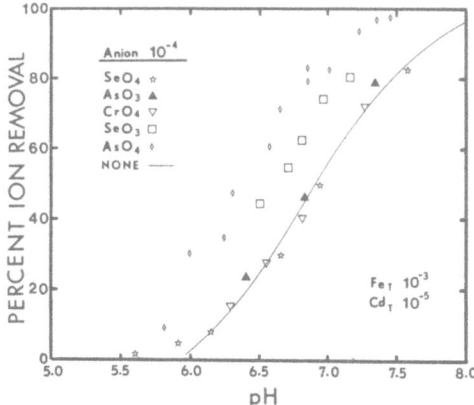

FIGURE 8: Effect of various anions on Cd adsorption for 10^{-4} M
anion added.

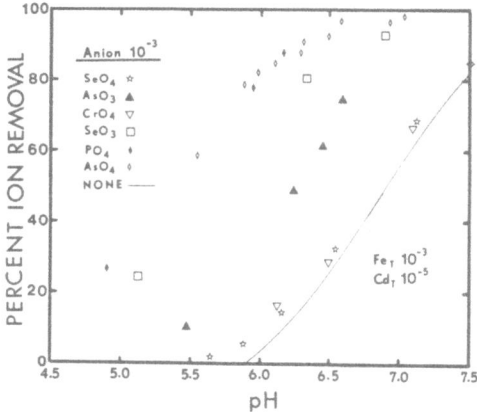

FIGURE 9: Effect of various anions on Cd adsorption for 10^{-3} M
anion added.

even when anion adsorption density attained its maximum limiting
value. The most obvious examples of this are the systems containing SeO_4^{2-}. As shown in Figure 3, fractional adsorption of SeO_4^{2-}
is limited to $\leq$ 20% at any pH in systems containing 10^{-3} M SeO_4^{2-}
and 10^{-3} M Fe. There are only two reasonable explanations for
this limitation. There may be an insufficient number of surface
sites to bind more than this amount of SeO_4^{2-}, in which case
competitive interactions with the metal would be expected.
Alternatively, SeO_4^{2-} adsorption may be limited electrostatically.
That is, adsorption of 20% of the total SeO_4^{2-} in solution may

Table 2. Acidity and speciation of the anionic adsorbates studied.

Anion Acid	Acidity constants			Speciation at pH 7.0*			
	$\log K_{a1}$	$\log K_{a2}$	$\log K_{a3}$	α_0	α_1	α_2	α_3
H_2SeO_3	2.3	7.9	–	$10^{-4.8}$	0.89	0.11	–
H_2SeO_4	<0	1.7	–	$<10^{-1.2}$	$10^{-5.3}$	1.0	–
H_2CrO_4	-0.1	6.5	–	$10^{-7.7}$	0.25	0.75	–
H_3AsO_3	9.3	12.2	?	0.99	0.01	$10^{-7.5}$	$<10^{-10.0}$
H_3AsO_4	2.2	7.0	11.5	$10^{-5.1}$	0.50	0.50	$10^{-4.8}$
H_3PO_4	2.2	7.2	12.3	$10^{-5.0}$	0.61	0.39	$10^{-5.7}$

$$ * \quad \alpha_1 \equiv \frac{(H_{n-i}A)^{-i}}{(A)_T} \ , $$ where H_nA is the uncharged acid and $(A)_T$ is the total concentration of A in the system.

make the surface charge sufficiently negative that additional SeO_4^{2-} adsorption is suppressed. However in this case adsorption of SeO_4^{2-} would significantly enhance cation adsorption. Thus, regardless of which factor limits SeO_4^{2-} adsorption, the effect of SeO_4^{2-} on metal adsorption should be significant; but no effect was observed, neither a competitive reduction in metal adsorption nor an electrostatic enhancement. It seems only remotely possible that the competitive and electrostatic interactions exactly balanced one another.

A second argument against a strong electrostatic effect of anion adsorption is based on the fact that the impact of anion adsorption on surface potential should be greatest for anions which form strong acids, and least for anions which form weak acids. This is because the strong acid anions are presumably the most highly charged adsorbed species. The acidity constants and speciation of these anions in solution at pH 7.0 are given in Table 2. The acidities increase in the order AsO_3^{3-} < SeO_3^{2-} < PO_4^{3-} < AsO_4^{3-} < CrO_4^{2-} < SeO_4^{2-}. The effects of the anions on cation adsorption bear no strong relation to their acidities. In fact, the strongest acid (H_2SeO_4) has the smallest effect. As noted earlier, the speciation of adsorbed ions may not correspond to that in solution. However, all the anions investigated are oxy-acids which probably bind to the surface through similar mechanisms. Based on the similarity in their solution phase reactions and in their adsorption behavior, it is reasonable to assume that these ions are similar in their surface speciation as well. While it is not likely that the speciation at the surface would be the same as that in solution for any given ion, the relative acidities of these anions probably follow the same sequence for adsorbed as for dissolved ions. This argument, along with the lack of any significant effect of SeO_4^{2-} or CrO_4^{2-} on cation adsorption, indicates that electrostatic interactions cannot be invoked as a primary explanation of the experimental results.

The combined effects of site competition and coulombic interactions can be simulated using the SGMA. In modeling the competitive systems, we have used the best fit stability constants from Table 1 for each adsorbing species. This involved using different stability constants for an adsorption reaction for different concentrations of adsorbate added. This choice was made because we wish any differences between the model and the experiments to result from competitive interactions, not from imperfect modeling of the adsorption behavior of the individual ions. When the constants determined in the single-adsorbate systems are used to model adsorption in the competitive systems, changes in surface potential due to anion adsorption are predicted to have a much greater effect on cation adsorption than competition for sites. That is, the presence of an anion always increases cation adsorption. The magnitude of the predicted effect is always much greater than that observed. Furthermore, in accord with the reasoning above, strong acids (e.g., SeO_4^{2-}, CrO_3^{-}) are predicted to have a greater effect than weaker ones (AsO_4^{3-}, SeO_3^{2-}) (Figures 10 and 11). As noted earlier, the observed order does not correlate with acidity. These modeling efforts show how one model can be used to make quantitative predictions about adsorption behavior in competitive systems. The SGMA was chosen for this purpose because it has been shown to reproduce both adsorption and electrophoretic mobility data in some systems.[7] While other models might predict different behavior, it is unlikely that any adsorption model could explain why the effects of the anions do not correlate with their acidities or why neither competitive nor electrostatic effects are apparent in certain systems containing large anion concentrations.

Differences between the predicted and observed results suggest that a separate factor may be dominating adsorption behavior of the metal to such an extent that electrostatic and competitive inter-actions are unimportant. One such possibility is that the anions form a new surface phase to which cations can adsorb. Since the order of increasing effect on metal adsorption is also the order

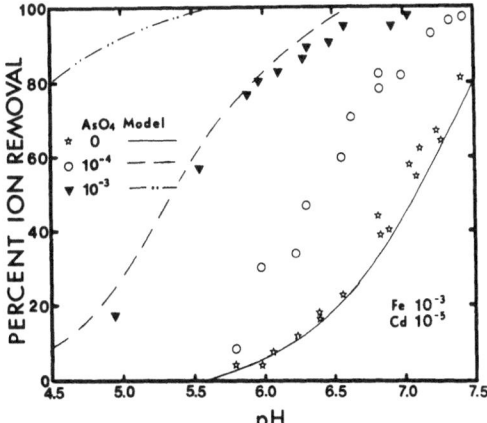

FIGURE 10:   Comparison of experimental results (data points)
                with predictions of the SGMA (smooth curves) for
                the effect of AsO_4^{3-} on Cd adsorption.

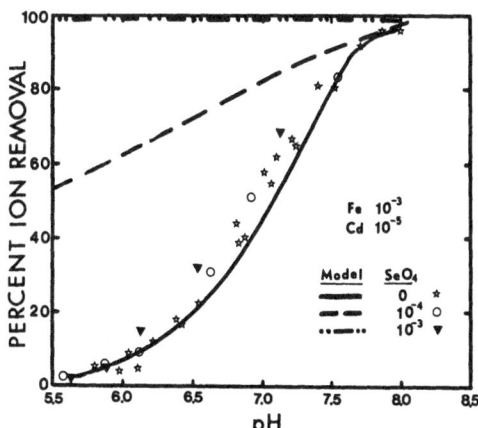

FIGURE 11:   Comparison of experimental results (data points)
                with predictions of the SGMA (smooth curves) for
                the effect of SeO_4^{2-} on Cd adsorption.

of decreasing solubility for most solids involving these anions,
this hypothesis would help explain the relative impacts of the
various anions on trace metal adsorption. For instance, phosphate
and arsenate have the greatest effects on trace metal removal
from solution and most metal phosphates and arsenates are very
insoluble. This result could not be interpreted in terms of
electrostatic interactions, since phosphate and arsenate do not
have significantly different acidities from the other anions

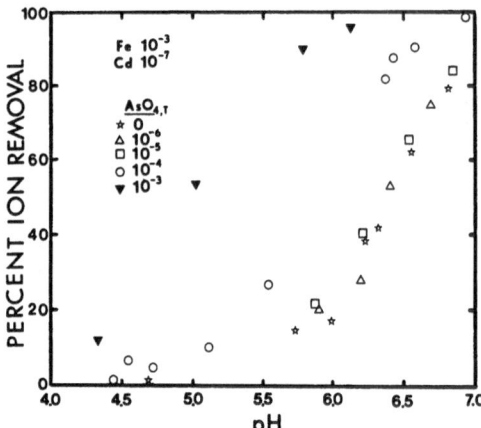

FIGURE 12:   Effect of AsO_4^{3-} on Cd adsorption in a system containing 10^{-7} M Cd_T. These results may be compared
with Figure 9, which represents similar systems
containing 10^{-5} M Cd_T.

studied. In addition, if cations bind to the new surface phase,
competition between anions and cations for sites on the original
substrate would be reduced or eliminated.

The anions may form a new surface phase with either the
trace metal or with iron atoms from the original substrate. To
distinguish these possibilities, the Cd-AsO$_4$ experiments were
repeated adding only 10^{-7} M Cd to solution, instead of 10^{-5} M Cd
(Figure 12). If precipitation of a Cd-AsO$_4$ solid was affecting
Cd behavior in these systems, then a much higher AsO_4^{3-} concentration would have been required in the 10^{-7} M Cd system, than in
the 10^{-5} M Cd system, for an equivalent effect. However, in both
cases, 10^{-5} M AsO_4^{3-} had no effect, 10^{-4} M AsO_4^{3-} enhanced Cd
removal somewhat, and 10^{-3} M AsO_4^{3-} caused a dramatic increase.
This suggests strongly that if a new surface phase is forming, it
must be comprised of the iron and arsenate. Further support for
this hypothesis is provided by the results of an experiment in
which 10^{-3} M AsO_4^{3-} was added to a 10^{-3} M Fe solution at pH 2.0.
Initially all the iron was soluble, but when the arsenate was
added, a white X-ray amorphous precipitate, presumably FeAsO$_4$,
formed. This suspension was then adjusted to pH 7.0, and Cd
adsorption onto the solid was studied. As shown in Figure 13,
Cd sorption onto FeAsO$_4$ is very similar to that on Fe$_2$O$_3\cdot$H$_2$O(am)
with AsO_4^{3-} adsorbed. While this result does not confirm the
presence of an FeAsO$_4$ surface phase in the latter system, it does
indicate that such a surface phase would have the observed effect
on Cd removal from solution.

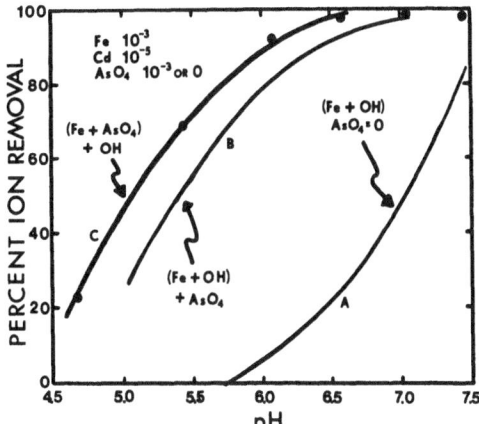

FIGURE 13: Comparison of Cd adsorption on $Fe_2O_3 \cdot H_2O$(am) (curve A) on $Fe_2O_3 \cdot H_2O$(am) with AsO_4^{3-} adsorbed (curve B) and on $FeAsO_4$ (curve C).

That formation of a new surface phase may accompany adsorption under some circumstances has been suggested by Tewari and Lee[18] based on X-ray photoelectron spectroscopic studies of the surface and by James and Healy[19] based on electrophoretic mobility studies. In both these cases, the adsorbate was a metal cation and the new surface phase was presumed to be a hydroxide of the adsorbed metal, while the new phases in our experiments apparently formed between surface iron atoms and adsorbing anions. Nevertheless the driving forces for precipitation would be the same in all the systems. Of particular note is James and Healy's derivation showing from theoretical considerations that the surface environment may induce precipitaton when the bulk solution is unsaturated.

Summarizing our hypothesis, at low surface coverages, anions adsorb and may alter the surface charge in accordance with one or more of the available adsorption models. However, at some surface density, depending on the identity of the anion and the surface potential, an iron-anion surface precipitate starts forming. At this point, either by surface diffusion or by sequential desorption/adsorption reactions, other adsorbed anions may adsorb to either the original (Fe-oxide) or new (Fe-anion) surface. If the new surface is a significantly stronger absorbent than the original, trace metal removal will increase dramatically when the new phase forms. If the new surface is a weaker adsorbent, trace metals will continue to bind to the original surface and be affected minimally by the precipitation reaction.

While this hypothesis resolves some of the questions raised earlier, the lack of any obvious anion-cation-surface interaction in some systems remains puzzling. In fact, if the surface precipitation hypothesis is valid, it may be that none of the anions affect cation adsorption through competitive or electrostatic interactions. In other words, if there were some way to force AsO_4^{3-} or PO_4^{3-} to adsorb without precipitating, there is no evidence at this time that they would affect metal ion adsorption any more so than SeO_4^{2-} or CrO_4^{2-} did. Two possible explanations of the observations are:

1. A new surface phase forms in all the systems investigated, but in some cases its sorption properties are identical to $Fe_2O_3 \cdot H_2O$(am).

2. Under conditions where no new surface phase forms, anions adsorb and alter the average surface potential. Trace metals bind at sites where the local electrical potential is different from the average potential, and is in fact nearly the same as in the system with no anions added. This requires that the metal-binding sites be specific for metals, i.e., anions cannot bind to them and therefore cannot reduce the number of sites available.

While neither possibility can be ruled out entirely, the first relies heavily on coincidence and is considered very unlikely. The possibility that anions and cations bind to separate groups of sites which are physically and electrically isolated from one another lacks independent supporting evidence and must also be considered speculative. Further experimentation to test this hypothesis is underway.

SUMMARY AND CONCLUSIONS

When strongly adsorbing anions and cations are present in a suspension, the ions may interact at the surface in a number of ways. There may be favorable or unfavorable electrostatic interactions, the various ions may compete for available surface sites, or some of the ions may form bridges between the surface and other ions. Adsorption onto $Fe_2O_3 \cdot H_2O$(am) of Cd, Zn, or Co in the presence of a strongly binding oxyanion (SeO_4^{2-}, SeO_3^{2-}, AsO_4^{3-}, AsO_3^{3-}, CrO_4^{2-}, or PO_4^{3-}) indicates that electrostatic and competitive interactions are unimportant. This result is contrary to predictions based on the SGMA. Only when the anions form a new surface phase with the iron, is cation adsorption affected significantly. Such a surface phase may form while the bulk solution is unsaturated. For all three trace metals studied, the impact of the various anions increases in the order $SeO_4^{2-} < AsO_3^{3-} \leq CrO_4^{2-} < SeO_3^{2-} \leq PO_4^{3-}$. The lack of significant electrostatic or competitive effects in any of the systems studied suggests that anions and cations may bind to groups of sites which are distinct and insulated from one another, but the causes of these distinctions are not known at this time.

ACKNOWLEDGEMENTS

The authors wish to thank Dr. J.F. Ferguson for his helpful comments. This work was supported by NFS Grant #7809056.

REFERENCES

1. J. Leckie and R. James, Control mechanisms for trace metals in natural waters, in: "Aquatic Environmental Chemistry of Metals", A.J. Rubin, ed., Ann Arbor Science Foundation, Ann Arbor, MI (1974).
2. L.J. Stryker and E. Matijevic, Adsorption of hydrolyzed hafnium ions on glass, in: "Adsorption from Aqueous Solution", W. Weber and E. Matijevic, eds., Advances in Chemistry Series #79, American Chemical Society, Washington, DC (1968).
3. T. Wakamatsu and D.W. Fuerstenau, The effect of hydrocarbon chain length on the adsorption of sulfonates at the solid/ water interface, in: "Adsorption from Aqueous Solution", W. Weber and E. Matijevic, eds., Advances in Chemistry Series #79, American Chemical Society, Washington, DC (1968).
4. E.A. Jenne, Controls on Mn, Fe, Co, Ni, Cu, and Zn concentrations in soils and water: the significant role of hydrous Mn and Fe oxides, in: "Adsorption from Aqueous Solution", W. Weber and E. Matijevic, eds., Advances in Chemistry Series #79, American Chemical Society, Washington, DC (1968).
5. M.A. Anderson, J.F. Ferguson and J. Gavis, J. Colloid Interface Sci. 54:391 (1976).
6. G.A. Parks and P.L. de Bruyn, J. Phys. Chem. 66:967 (1962).
7. J.A. Davis, R.O. James and J.O. Leckie, J. Colloid Interface Sci. (63(3):480 (1978).
8. W. Stumm, C. Huang and S. Jenkins, Croatica Chim. Acta 43:223 (1970).
9. R. James and T. Healy, J. Colloid Interface Sci. 40:65 (1972).
10. J.A. Davis and J.O. Leckie, Speciation of adsorbed ions at the oxide/water interface, in: Chemical Modeling in Aqueous Systems", E.A. Jenne, ed., Advances in Chemistry Series #93, American Chemical Society, Washington, DC (1979).
11. H. Hohl and W. Stumm, J. Colloid Interface Sci. 55:281 (1976).
12. M.M. Benjamin and J.O. Leckie, J. Colloid Interface Sci. accepted for publication.

13. J.A. Davis and J.O. Leckie, J. Colloid Interface Sci. 67:90 (1978).

14. J.A. Davis and J.O. Leckie, J. Colloid Interface Sci. 74:32 (1980).

15. M.M. Benjamin and J.O. Leckie, Competitive adsorption of Cd, Cu, Zn, and Pb on amorphous iron oxyhydroxides, submitted for publication.

16. J.O. Leckie, M.M. Benjamin, K. Hayes, G. Kaufman and S. Altmann, "Adsorption/Coprecipitation of Trace Elements from Water with Iron Oxyhydroxide", prepared for the Electric Power Research Institute as EPRI-RP-910 (1979).

17. M.L. Pierce and C.B. Moore, Environ. Sci. and Technol. 14:214 (1980).

18. P.H. Tewari and W. Lee, J. Colloid Interface Sci. 52:77 (1975).

19. R.O. James and T.W. Healy, J. Colloid Interface Sci. 40:53 (1972).

SPECIFIC ADSORPTION OF Co(II) AND

[Co(III)EDTA]⁻ COMPLEXES ON HYDROUS OXIDE SURFACES

C.P. Huang and Y.T. Lin

Environmental Engineering Program
Department of Civil Engineering
University of Delaware, Newark, Delaware 19711

ABSTRACT

The effect of complex formation on the adsorption behavior
of heavy metals is studied by comparing the adsorption characteristics of Co(II) species, (uncomplexed), and [Co(III)EDTA]⁻
complexes on two hydrous solids, namely SiO_2 and Al_2O_3. The
results demonstrate that for highly stable metal complexes such
as [Co(III)EDTA]⁻, and particularly ones having three-dimensional
structure, their adsorption behavior is influenced by the nature
of their ligands. However, their adsorption characteristics do
not necessarily copy that of the complex-forming agent. Furthermore, specific chemical interactions with the formation of hydrogen bondings play a very significant role in the adsorption of
both the complexed and uncomplexed metal ions on hydrous solids.
Adsorption reactions between metal ions, both complexed and uncomplexed, and hydrous solids can be readily described by a surface
complex formation model in that the surface hydroxo groups function
as the ligands.

INTRODUCTION

Adsorption of metal ions onto hydrous solid surfaces is an
important interfacial reaction that has multiple industrial
applications, e.g. mineral processing and pollution control.
Much research effort has been expended on the adsorption of metal
ions onto various hydrous solids.[1-10] Of particular importance
relating to metal adsorption on hydrous solids is the coexistence
in the system of complex-forming agents (or chelates) and metal
ions. Complexation is always accompanied by changes in adsorption-

related properties of the metal, such as oxidation state, chemical
compatibility with the solvent, i.e. solubility, extent of solvation, molecular size and stereochemical configuration.

The effect of complex formation on the adsorption characteristics of metal ions is not entirely understood, although it is
obvious that there will be three possible outcomes, i.e. enhancement, inhibition or no effect. Vuceta[11] suggested that the
addition of ligand may i) inhibit adsorption because of strong
complex formation with the metal ions or by competing with the
surface for available adsorption sites; ii) cause no perceptible
change if the ligand exhibits weak complexation ability and a
lack of affinity for the solid surface, or iii) enhance adsorption
if the ligand is capable of strong complex formation and also
possesses a substantial affinity for the solid surface. This is
in accord with what was suggested by Benjamin,[12] that there are
at least three possible ways by which the metal-ligand complex
may interact with the surface: i) adsorption of complexed metal
ions depends upon the adsorbability of uncomplexed metal (metal-
like adsorption); ii) adsorption of the complexed metal ions is
related to the adsorption characteristics of the free ligand
(ligand-type adsorption); and iii) the complexed metal ions may
not be adsorbed at all. A recent review on the effect of complex
formation on the adsorption characteristics of heavy metals
indicated that difficulties still remain in generalizing about
the influence of chelates on metal adsorption.[13] Ligand/metal
ratio, adsorbability of the free ligand, and pertinent solution
chemistry (e.g. pH) are but some of the many factors acting
collectively or independently on the extent of adsorption of
metals in the presence of chelating agents. They further asserted
that the effect of complex formation on metal adsorption must be
assessed by the adsorbability of the complexed metal ions. This
observation was further supported by their recent work on the
adsorption characteristics of Cu(II) in the presence of chelating
agents.[14,15]

The objectives of the present work were to examine the
effect of complex formation on the adsorption characteristics of
Co(II) and to gain insight into the mode of the adsorption of
Co(III) complexes on the hydrous solids. Since Co(III) complexes
are always octahedral, they represent an ideal stereochemical
system for investigation.

In aqueous solutions, cobalt is found in two possible oxidation states: Co(II) and Co(III). Without the presence of strong
ligands, Co(III) can oxidize water molecules and is therefore
unstable. However, when strong ligands are present, Co(III)[16]
becomes more stable than Co(II) due to ligand field effects.
For instance, $Co(NH_3)_6^{3+}$ is more stable than $Co(H_2O)_6^{2+}$ in aqueous
solutions. Strong ligands such as ethylenediaminetetraacetate

(EDTA) can form a very strong Co(III) complex ($K_1 = 10^{36}$) which
is much stronger than [Co(II)EDTA]$^{2-}$ ($K_1 = 10^{16}$). In homogeneous
aqueous solutions, hydrated [Co(III)EDTA(H$_2$O)]$^-$ species are not
stable, being slowly dehydrated to [Co(III)EDTA]$^-$ complexes.
Shimi and Higginson report that it takes from five hours (at
pH 4 – 8) to five days (pH > 9 or pH < 4) to achieve 99.9% of
conversion equilibrium.[17] Moreover, [Co(III)EDTA(H$_2$O)]$^-$ behaves
as an amphoteric acid and base in aqueous solution, with acidity
constants $pK_{a1} = 3$ and $pK_{a2} = 8$:

$$\left[Co(III)HY(H_2O)\right] \underset{+H^+}{\overset{-H^+}{\rightleftharpoons}} \left[Co(III)Y(H_2O)\right]^- \underset{+H^+}{\overset{-H^+}{\rightleftharpoons}} \left[Co(III)Y(OH)\right]^{2-} \quad (1)$$

where Y stands for EDTA.

Table 1 shows the various equilibria possible with Co(II) and
Co(III) species and the corresponding equilibrium constants.[17,18]

MATERIAL AND METHODS

Chemicals

Cobaltous chloride, disodium ethylenediaminetetraacetate (EDTA),
hydrogen peroxide and sodium perchlorate were purchased from the
Fisher Scientific Company, Fair Lawn, New Jersey. Cobaltous
chloride isotope (99% ^{60}Co) was purchased from New England Nuclear
Co., Boston, Massachusetts. Scintillation cocktail solution
(Beckman HP) was purchased from Beckman Instruments Co., Fuller,
California. Solids, Aluminum Oxide C and Cabosil, were obtained
respectively from Degussa, Inc., New York and Cabot Co., Boston,
Massachusetts. Aluminum Oxide C is a fine powder of primary
γ-Al$_2$O$_3$ structure; whereas Cabosil is an ultrafine silica (SiO$_2$)
made by flame hydrolysis process. Both solids have very high
purities ($\geqslant 99\%$) and were used as received without further treatment. Table 2 summarizes the typical properties of these two
solids.

Distilled and deionized water were used throughout the whole
experimentation.

Adsorption of Co(II)

Batch adsorption experiments were carried out by mixing
appropriate amounts of stock Co(II) solution (10^{-3} M CoCl$_2$ + 10^{-1} M
HClO$_4$), stock H$_2$O$_2$ solution (30%), stock solid suspensions (25 g/L)
and stock ionic medium (0.5 M NaClO$_4$) to give an overall concentra-

Table 1.   Chemical Equilibria in Co(II) EDTA H_2O Systems

Equilibria	K	Log K	Ref.
$Co(OH)_2(s) \rightleftharpoons Co^{2+} + 2OH^-$;	K_{so}	-14.90	(18)
$Co^{2+} + H_2O \rightleftharpoons Co(OH)^+ + H^+$;	$*K_1$	- 9.60	(18)
$Co(OH)^+ + H_2O \rightleftharpoons Co(OH)_2 + H^+$;	$*K_2$	- 9.20	(18)
$Co(OH)_2(s) \rightleftharpoons Co(OH)_2$;	K_{s2}	- 5.70	(18)
$Co(OH)_2 + H_2O \rightleftharpoons Co(OH)_3^- + H^+$;	$*K_3$	-12.00	(18)
$Co^{2+} + EDTA^{4-} \rightleftharpoons CoEDTA^{2-}$;	K_1	16.20	(18)
$Co(OH)_3(s) \rightleftharpoons Co^{3+} + 3OH^-$;	K_{so}	-25.60	(18)
$Co(OH)_3(s) \rightleftharpoons Co(OH)_3(aq)$	K_{s3}	- 4.54	(18)
$Co^{3+} + H_2O \rightleftharpoons Co(OH)_2^{2+} + H^+$;	$*K_1$	- 1.80	(18)
$Co^{3+} + Y^{4-} \rightleftharpoons CoY^{-1}$;	K_1	36.00	(18)
$CoY^- + H_2O \rightleftharpoons Co(H_2O)Y^-$;	K_h	< 1.00(?)	(17)
$Co(H_2O)Y^- \rightleftharpoons Co(OH)Y^{2-} + H^+$;	K_{a2}	- 8.10	(17)
$Co(OH_2)HY^+ \rightleftharpoons Co(OH_2)Y + H^+$;	K_{a1}	- 3.10	(17)

$Y = EDTA^{4-}$

Table 2.   Properties of Oxide Particles

Solid	Particle Size (μm)	Specific Surface Area (M^2/g)	pH_{zpc}
γ-Al$_2$O$_3$	0.02[a]	100 ± 15[b]	8.45[c]
SiO$_2$ (EH-5) (am)	0.007[a]	325 ± 40[b]	3.5 ∿ 4.2[c]

a. Data provided by the manufacturer
b. Determined in this laboratory by N$_2$-gas
    adsorption with a BET apparatus, Monosorb.
    Quantachrome, N.Y.
c. Determined in this laboratory by electrophoretic
    mobility measurement.

tion of 6 x 10⁻⁴ M Co(II), 2.5 g/L solid and 0.025 M NaClO$_4$ with a final total solution volume of 100 mL (in 125 mL polyethylene bottles). In a separate experiment, H$_2$O$_2$ was added to the above Co(II) solution at an overall concentration of 0.4 M. It was expected that the H$_2$O$_2$ present at an excess quantity may oxidize Co(II) to Co(III) to allow the observation of Co(III) adsorption.

The pH values were first adjusted with strong acid, HClO$_4$ and/or strong base NaOH to cover a range from 3 to 10. After pH adjustment, the mixtures were shaken in a shaking machine for eight hours at ambient temperature (25 ± 1° C). At the end of the shaking time, the final pH value was measured and recorded with an Orion model 810A pH-meter equipped with a combination electrode. The suspensions were then centrifuged at 15,000 rpm (250,000 g) for 15 minutes with a DuPont Sorvall Temperature-controlled centrifuge. The residual concentration of Co(II), was determined as its EDTA complex, [Co(III)EDTA]⁻ at 540 nm with a Hitachi-Perkin Elmer spectrophotometer. Twenty-five mL of the supernatant was pipetted into a 50-mL bottle containing 6% H$_2$O$_2$ as an oxidizing agent. To insure complete formation of [Co(III)EDTA]⁻ complex, the mixture was allowed to stand for four days before spectrophotometric analysis.

Adsorption of EDTA

The procedure for EDTA adsorption experiments followed that of the Co(II) adsorption. The overall EDTA concentrations varied from 1 x 10⁻⁴ M to 8 x 10⁻⁴ M. A modification of Kaiser's[25] method was used to determine the residual concentration of EDTA. In this, 25 mL of supernatant was first pipetted into a 50 mL volumetric flask, followed by the addition of 2 mL of 10^{-2} M $CoCl_2$, 1 mL of 200 gm/L $NaNO_2$, 2 mL of 0.5 M NaF, and 0.3 mL of 3% HCl. The mixtures were hand-shaken for 30 seconds then set aside for three minutes before adding 2 mL of 6% H_2O_2. The flask was then filled to the mark with distilled-deionized water, mixed well and measured for absorbance at 540 nm using a 10 cm optical cell with a Hitachi-Perkin Elmer spectrophotometer.

Adsorption of [Co(III)EDTA]⁻

A stock solution of [Co(III)EDTA]⁻ complex was prepared by oxidizing a Co(II) solution containing equimolar concentration of EDTA with H_2O_2. To an 800 mL solution containing 3.7224 g Na_2EDTA and 2.3793 g $CoCl_2$, was added 10 mL of 30% H_2O_2 three successive times over a 24 hour interval while stirring. The solution was then diluted with distilled-deionized water to the 1000 mL mark.

The experimental procedure for [Co(III)EDTA]⁻ adsorption was essentially the same as above for Co(II) and EDTA adsorption.

Resembling $KMnO_4$ in color, [Co(III)EDTA]⁻ solution appears purple at high concentrations and turns pink after dilution. The dilute [Co(III)EDTA]⁻ solution obeys the Beer's Law up to a concentration of 4×10^{-4} M (with 10 cm optical cell) at a wave length of 540 nm.

When ^{60}Co was used for adsorption experiments, the experimental procedures were essentially the same as above except that the percentage of Co(II)/or [Co(III)EDTA]⁻ adsorbed was measured by comparing the initial and the residual radioactivities (from β-decay) in the supernatant. Experimentally, 1 mL of ^{60}Co containing supernatant was mixed with 5 mL of Beckman Ready-Solve[R] HP cocktail liquid; the mixture was then placed in a Beckman 81,000 Scintillation Counter for β-particle counting with wide-opened window.

SURFACE ACIDITY OF HYDROUS SOLIDS

The surface acidity of γ-Al_2O_3 has been determined by Huang and Stumm[19] and recently by Hohl and Stumm[1] using alkalimetric titration technique. From the titration curve, two distinct regions of proton transfer can be recognized for γ-Al_2O_3. Thus the hydrous γ-Al_2O_3 surface can be treated as a diprotic solid Brönsted acid:

$$\equiv AlOH_2^+ \rightleftharpoons \equiv AlOH + H^+; K_{a1}(int) \tag{2}$$

for which

$$K_{a1}(int) = \frac{\{\equiv AlOH\}\{H^+\}}{\{\equiv AlOH_2^+\}}$$

and

$$\equiv AlOH \rightleftharpoons \equiv AlO^- + H^+; K_{a2}(int) \tag{2a}$$

for which

$$K_{a2}(int) = \frac{\{\equiv AlO^-\}\{H^+\}}{\{\equiv AlOH\}}$$

where $\{i\}$ designates the surface concentration (mol/g) of the ith species.

The total standard free energy change for the deprotonation process, which is related to K_a' is the sum of the change of free energy for H^+ dissociation from its surface site (related to $K_{a1}(int)$ or $K_{a2}(int)$) and the change of free energy for the transport of the dissociated proton away from the surface into the bulk solution as expressed by the Boltzman (or electrostatic) factor:

$$K_a' = K_a(int) \exp (F\psi_s/RT)$$

where F, ψ_s and RT are respectively, Faraday's constant, surface potential and the product of the gas constant and the absolute temperature.

The intrinsic constants and total number of surface hydroxo sites, Al_T, of γ-Al_2O_3 have been determined previously. Huang and Stumm[19] reported 7.9 and 9.1 for $pK_{a1}(int)$ and $pK_{a2}(int)$, respectively and 1.25×10^{-4} mol/gm (or 133 Å²/site) for Al_T. Hohl and Stumm[1] gave similar intrinsic constants 7.0 for $pK_{a1}(int)$ and 9.5 for $pK_{a2}(int)$. However, their Al_T value was slightly different from that of Huang and Stumm, at 2.1×10^{-4} mol/g (or 79 Å²/sites).[19]

The fractions of surface hydroxo groups, i.e. α_0, α_1, and α_2 for $\equiv AlOH_2^+$, $\equiv AlOH$, and $\equiv AlO^-$, respectively are calculated by the following relationships:

$$\alpha_0 = \frac{\{\equiv AlOH_2^+\}}{Al_T} = \frac{[H^+]^2}{K'_{a1}K'_{a2} + K'_{a1}[H^+] + [H^+]^2} \qquad (3a)$$

$$\alpha_1 = \frac{\{\equiv AlOH\}}{Al_T} = \frac{K_{a1}[H^+]}{K'_{a1}K'_{a2} + K'_{a1}[H^+] + [H^+]^2} \qquad (3b)$$

$$\alpha_2 = \frac{\{\equiv AlO^-\}}{Al_T} = \frac{K'_{a1}K'_{a2}}{K'_{a1}K'_{a2} + K'_{a1}[H^+] + [H^+]^2} \qquad (3c)$$

Where $[H^+]$ is the bulk proton concentration and K'_{a1} and K'_{a2} are the microscopic acidity constants. Figure 1a, adopted from Stumm et al.,[20] shows the molar distribution of the surface hydroxo groups of γ-Al_2O_3 as a function of solution pH.

The surface acidity of various brands of SiO_2 forms has been determined by Schindler,[21,22] Bolt,[23] and Huang.[24] Due to the low pH_{zpc} of most $SiO_2(s)$ materials, only the second acidity constant pK_{a2}(int) or K'_{a2} is attainable. The speciation of surface hydroxo groups for the $SiO_2(s)$ used in this study is shown in Figure 1b.[24] The values Si_T and pK_{a2}(int) are 2.6 x 10^{-4} mol/g (or 208 Å^2/site) and 6.7, respectively.

RESULTS

The adsorption of Co(II), in the absence of EDTA, onto γ-Al_2O_3 and SiO_2 surfaces was studied as a function of suspension pH at various surface loadings, σ_o, (i.e. the maximum adsorption density that will be possible if all adsorbates present were adsorbed). For both solids, the amounts of Co(II) adsorbed increase abruptly at a specific pH value, namely 6-7. When H_2O_2 was present, the metal species adsorbed was expected to be Co(III) ions. The results shown in Figure 2(a) indicate very little difference in Co(II) adsorption onto the hydrous oxide surfaces whether H_2O_2 is present or not. The results also demonstrate that the pH of abrupt adsorption shifts from 6 to 7 when the surface loading increases from 9.2 x 10^{-13} mol/g to 2.4 x 10^{-4} mol/g. The results presented in Figure 2 agree well with what was reported by others.[1-5]

Figure 3 shows the adsorption of EDTA species onto hydrous solids. For γ-Al_2O_3, the adsorption density (mol EDTA adsorbed per gm solid) decreases as pH increases. The amount of EDTA adsorbed is negligible at pH > 10, a coincidence with the pK_{a4} of

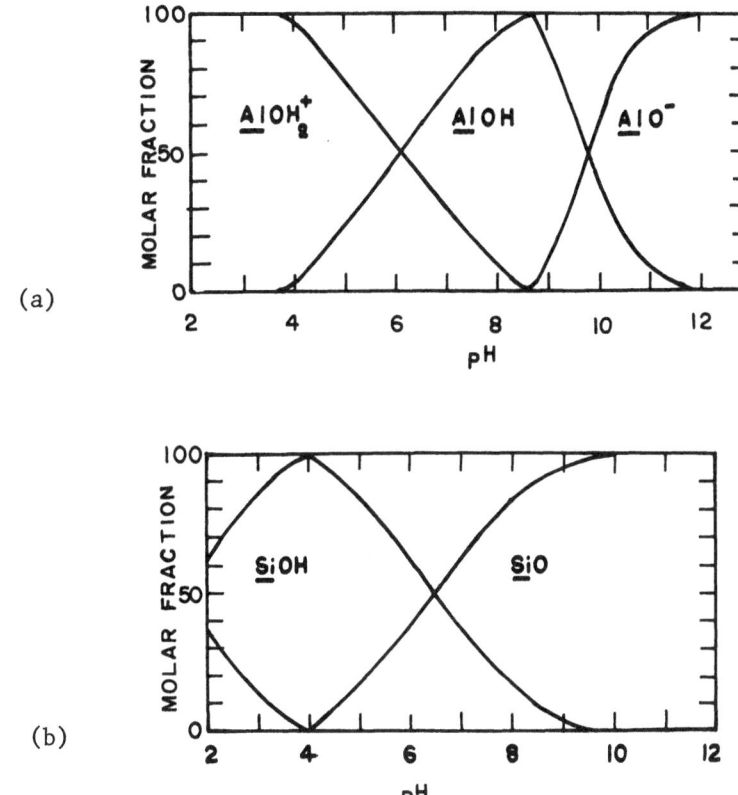

FIGURE 1: Distribution of surface hydroxo groups (a) for γ-Al₂O₃ and (b) for SiO₂. Figure 1a was adopted from Stumm et al.[20]

EDTA (pK_{a1} = 2.18, pK_{a2} = 2.73, pK_{a3} = 6.20, pk_{a4} = 10.0).[18] This observation agrees with what was recently reported by Rubio and Matijevic.[26] The degree of EDTA adsorption on SiO₂ surface is insignificant. Electrostatic repulsion seemingly prevents the anionic EDTA species from approaching the highly negatively charged SiO₂ surfaces. Moreover, the prevalence of deprotonated SiO₂ surface seemingly disallows the formation of specific chemical bonds such as hydrogen bonding. The results also indicate that negative adsorption becomes plausible as EDTA concentration decreases.

The adsorption density of [Co(III)EDTA]⁻ complex onto γ-Al₂O₃ and SiO₂ surfaces, at various surface loadings, as affected by pH is shown in Figure 4. For γ-Al₂O₃, three adsorption regions are

(a)

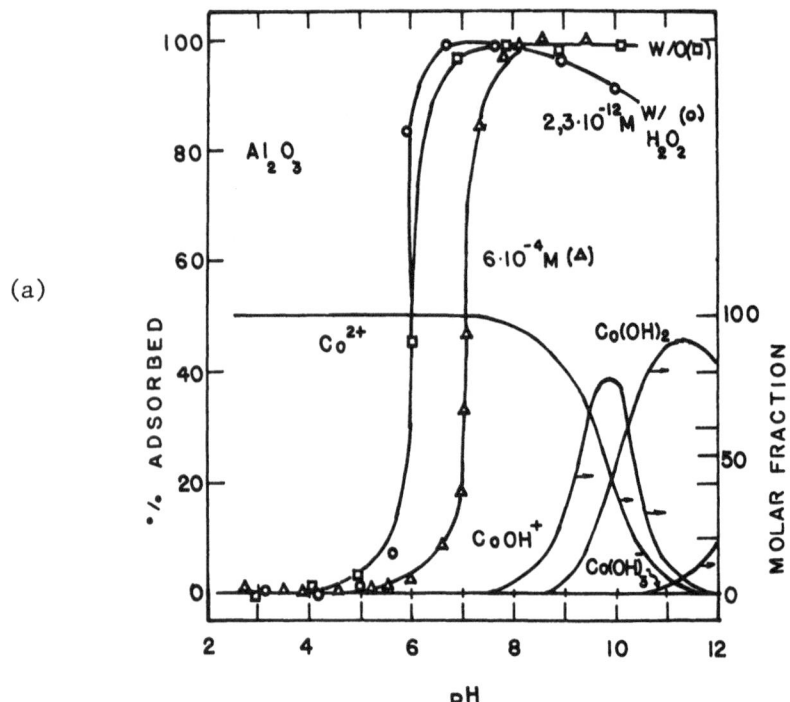

FIGURE 2: Adsorption characteristics of Co(II) on hydrous solids.
(a): open circles represent adsorption (%) from solution containing 2.3 x 10^{-12} M [Co(III)-EDTA]$^-$, 2.5 g/L
Al_2O_3 and 0.4 M H_2O_2. Open squares are adsorption from
the same solution as above without H_2O_2. The open
triangles are adsorption experiments conducted with
spectrophotometric method for a solution containing a
high Co(II) concentration at 6 x 10^{-4} M and 2.5 g/L
Al_2O_3. The speciation of Co(II) species is also shown.
In (b) the open diamonds are obtained from spectrophotometric measurement with 6 x 10^{-4} M Co(II) and 2.5 g/L
SiO_2. The solid dots were taken from James and Healy[3]
for comparison. Total Co(II) concentration was
1.2 x 10^{-6} M and total solid concentration was 75 m^2/L
(or 25 g/L). The open squares and triangles are adsorption from 2.3 x 10^{-12} M Co(II) and 25 g/L without and
with 0.4 M H_2O_2, respectively. The ionic strength was
0.05 M as $NaClO_4$ for both parts.

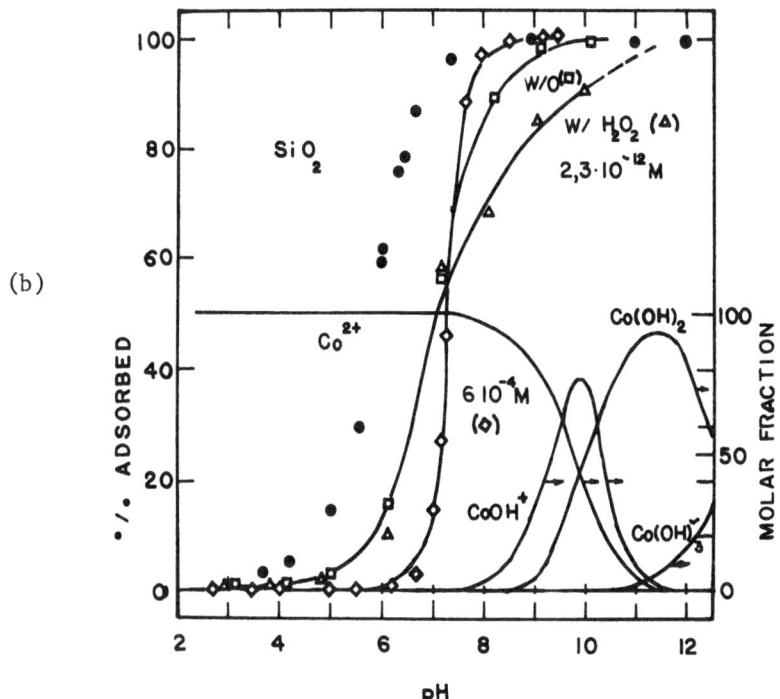

(b)

readily recognized: region I (pH < 5.5 ± 0.25), region II
(5.5 ± 0.25 < pH < 7.25 ± 0.25) and region III (pH ≥ 7.25 ± 0.25).
In region I, adsorption density decreases, almost linearly, with
increasing pH; in region II, the amount adsorbed increases rather
moderately with increasing pH to a peak value then decreases
rapidly with further increase in pH, and in region III, the
adsorption density increases asymptotically with increasing pH
values to approximately 10.

The three adsorption regions shown in Figure 4a which were
obtained with the spectrophotometric method were also observed in
a separate experiment using a ^{60}Co isotope (Figure 6a).

The extent of [Co(III)EDTA]⁻ adsorption on SiO₂ differs
significantly from that of γ-Al₂O₃. Results obtained from spectrophotometric measurement show that adsorption density increases
with increasing pH to a maximum adsorption peak then decreases to
a negligible level before markedly increasing at pH 8.5. However,
the adsorption peak between pH 5 and 8.5 was not observed when
the [Co(III)EDTA]⁻ complex was labeled with ^{60}Co isotope (Figure 6b).
Therefore, this adsorption peak is a manifestation of the spectrophotometric technique.

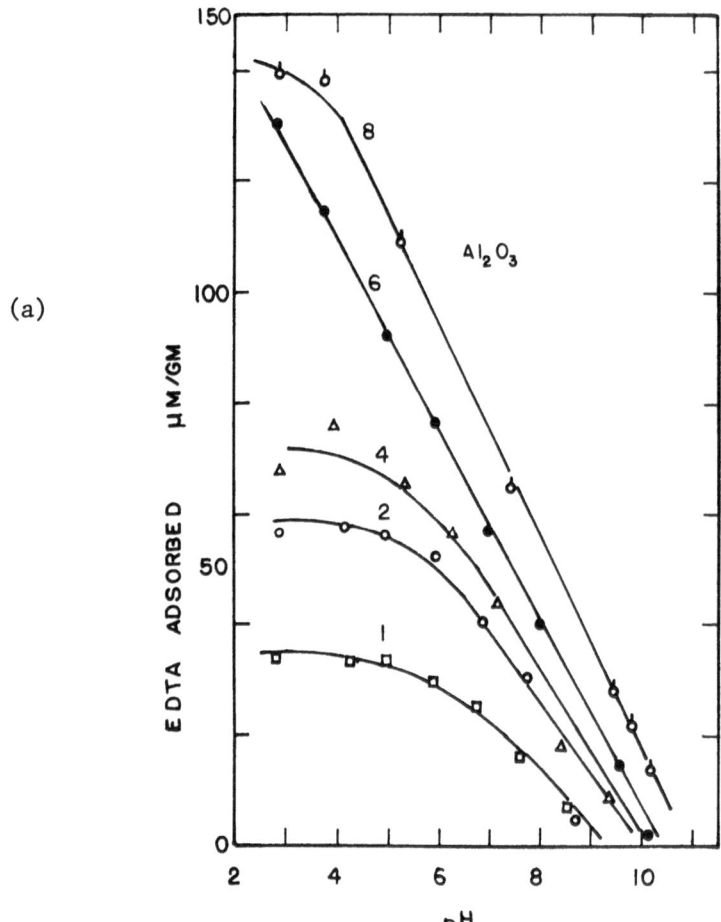

(a)

FIGURE 3:   The adsorption characteristics of EDTA on hydrous solids.
            In (a), the total solid concentration was 2.5 g/L
            ionic strength was 0.05 M as NaClO4.   The digits near
            each curve represent the total EDTA concentration in
            10^{-4} M.   In (b), the ionic strength and the solid
            concentration are the same as above.   The digit gives
            the total concentrations of EDTA in 10^{-4} M.

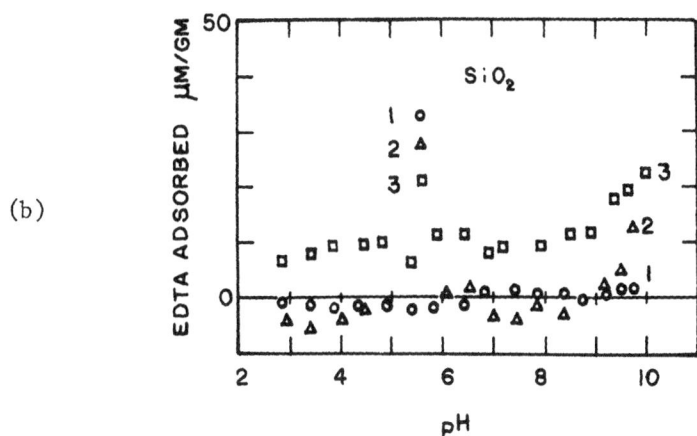

(b)

DISCUSSION

Adsorption of Co(II)

Many models have been developed to describe the adsorption characteristics of aqueous cations onto hydrous solid surfaces. These include the ion-exchange; the surface complex formation, solvation and specific adsorption models.[1] In this discussion, the surface complex formation concept is employed.

For γ-Al$_2$O$_3$ surface, it appears that reaction between Co^{2+} ions and the neutral surface hydroxo group, $\equiv$AlOH, with the formation of a surface complex $\equiv$AlOCo$^+$ or,

$$Al\!\!\!-\!\!\!\!-\!\!\! O \!\!-\!\!\! \overset{+}{Co} \!\!-\!\!\! \qquad (a)$$

is the major adsorption scheme:

$$Co^{2+} + \equiv AlOH \rightleftharpoons \equiv AlOCo^+ + H^+; \; {}^*K_1^s \qquad (4)$$

Although the deprotonated surfaces, dominated by $\equiv$AlO⁻ groups, have the electrostatic advantages in attracting Co^{2+} species, it appears not to be directly involved in the adsorption reaction, at least at pH values less than 8.2, the pH value at which 100% of Co(II) adsorption occurs, since at pH < 8.2 the oxide surface is protonated, (Figure 1a). For the same reason, the contribution to

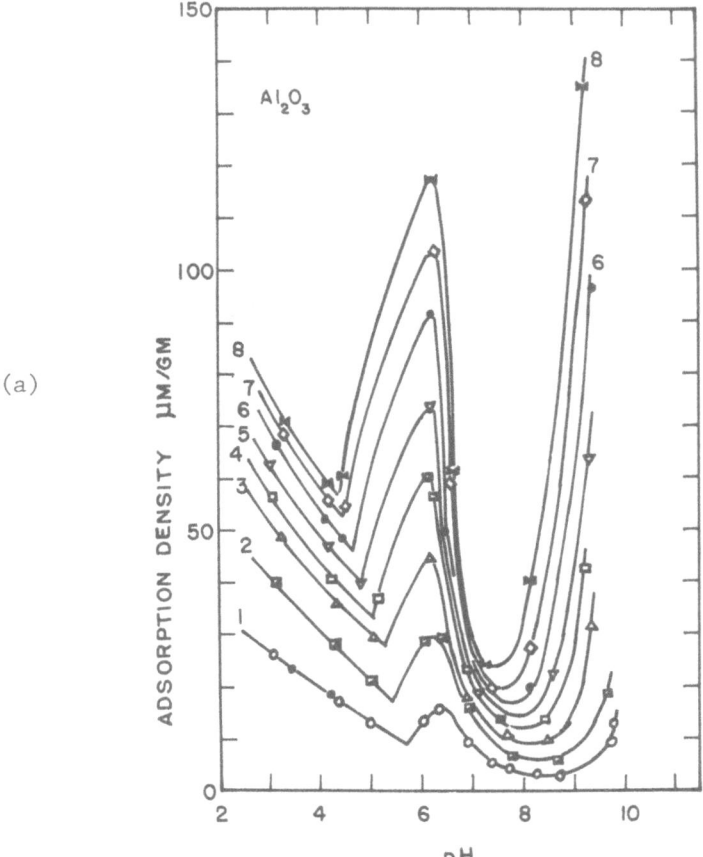

FIGURE 4: The adsorption characteristics of [Co(III)EDTA] com-
           plexes on hydrous solids. The total solid concentra-
           tion and ionic strength are at 2.5 g/L of (a) Al_2O_3
           and (b) SiO_2 respectively. The digits stand for
           total [Co(III)EDTA]$^-$ concentration in 10^{-4} M. It
           is to be noted that the peaks between pH 5.5 and 8.5
           for SiO_2 are not due to real adsorption, rather they
           resulted from reduction of absorbance due to the
           formation of hydrated [Co(III)EDTA(H_2O)]$^-$ complex.

Co(II) adsorption from reaction between $CoOH^+$ ions and $\equiv AlOH$
cannot be too significant, since at pH < 8.2 although there are
$\equiv AlOH$ groups available, the concentration of $CoOH^+$ in the solution
is negligible.

    The calculated $*K_1^S$ values remain relatively steady over a
wide pH range: $10^{-3.1}$ (pH 5) to $10^{-2.6}$ (pH 8.2). This indicates
that electrostatic interaction plays a rather insignificant role
in the Co(II) adsorption. Equilibrium constants based upon the

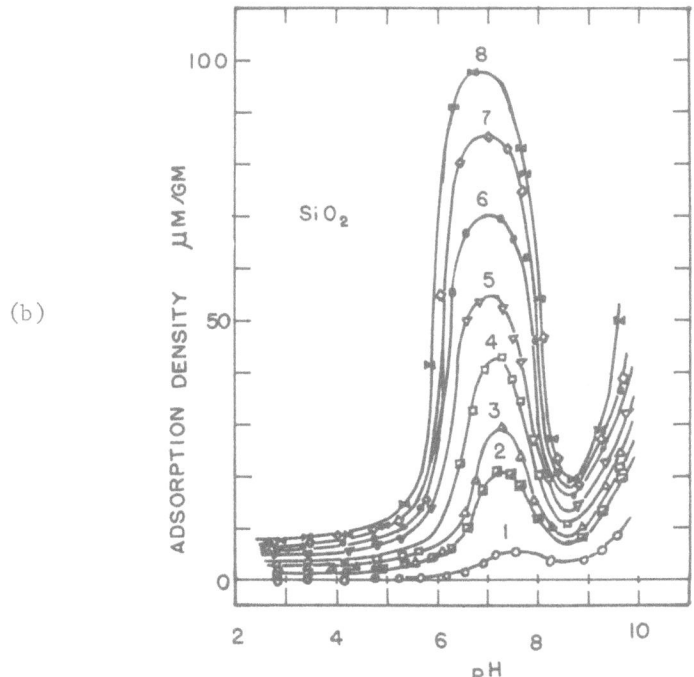

ADSORPTION DENSITY μM/GM

SiO₂

(b)

pH

following assumed reaction scheme are not as invariant as $*K_1^s$ over the pH range of interest.

$$Co^{2+} + 2\ \equiv\!AlOH \rightleftharpoons (\equiv\!AlO)_2\ Co + 2H^+, \ *\beta_2^s \tag{5}$$

The calculated $*\beta_2^s$ values vary from $10^{-5.6}$ (pH 5) to $10^{-11.1}$ (pH 8.2). This implies that the surface complex $(\equiv\!AlO)_2Co$, or

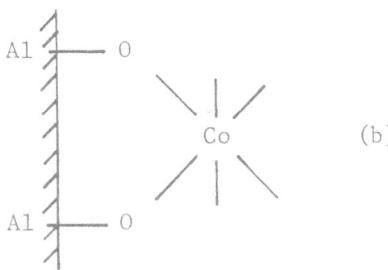

is not likely to be a major one. This is in contrast with what was suggested by Dalang and Stumm[2] who reported that the adsorption of Pb(II) on γ-Al₂O₃ can be described by a combination of the above

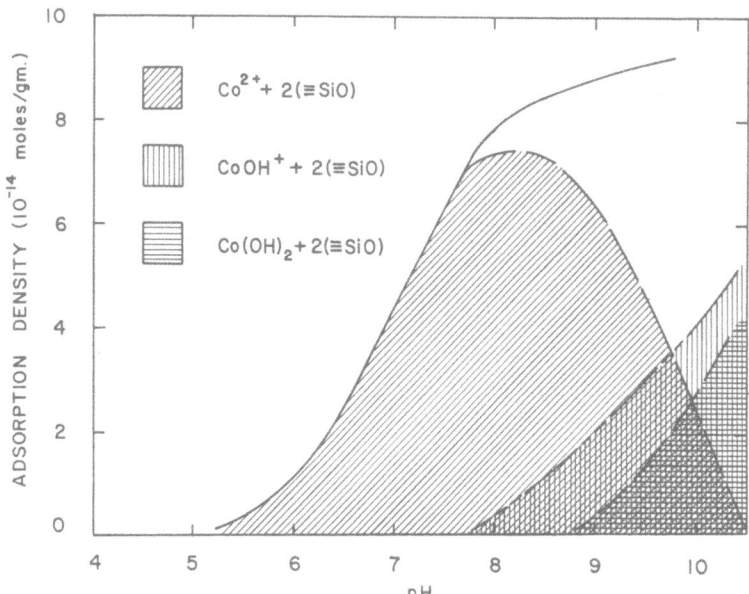

FIGURE 5: Formation of surface complexes in Co(II)-SiO$_2$ system.
The adsorption density contributed from the three
individual reaction schemes add up to the total amount
of Co(II) adsorbed on SiO$_2$. The model calculation
was based upon the results presented in Figure 2b
(open squares).

two reaction schemes. Although the average values of p*K$_1^S$ and
p*β$_2^S$ for Co(II) adsorption on γ-Al$_2$O$_3$ are in agreement, in order of
magnitude, with what was reported by Hohl and Stumm for Pb(II)
adsorption on the same hydrous solid, i.e. 3.3 and 7.5 versus 2.2
and 8.1 respectively for Co(II) and Pb(II). The results obtained
in this study tend to indicate that the adsorption of Co(II) on
hydrous γ-Al$_2$O$_3$ can be best described by Equation 4.

The adsorption of Co(II) on SiO$_2$ cannot be quantitatively described by either Equation 4 or Equation 5. Rather, the following
reaction schemes are suggested:

$$Co^{2+} + 2\equiv SiO^- \rightleftharpoons (\equiv SiO)_2 Co; \; \beta_2^S \tag{6}$$

$$CoOH^+ + 2\equiv SiO^- \rightleftharpoons (\equiv SiO)_2 CoOH^-; \; \beta_{1,2}^S \tag{7}$$

$$Co(OH)_2 + 2\equiv SiO^- \rightleftharpoons (\equiv SiO)_2 Co(OH)_2^{2-}; \; \beta_{2,2}^S \tag{8}$$

The above reactions lead to the formation of the following surface
complexes

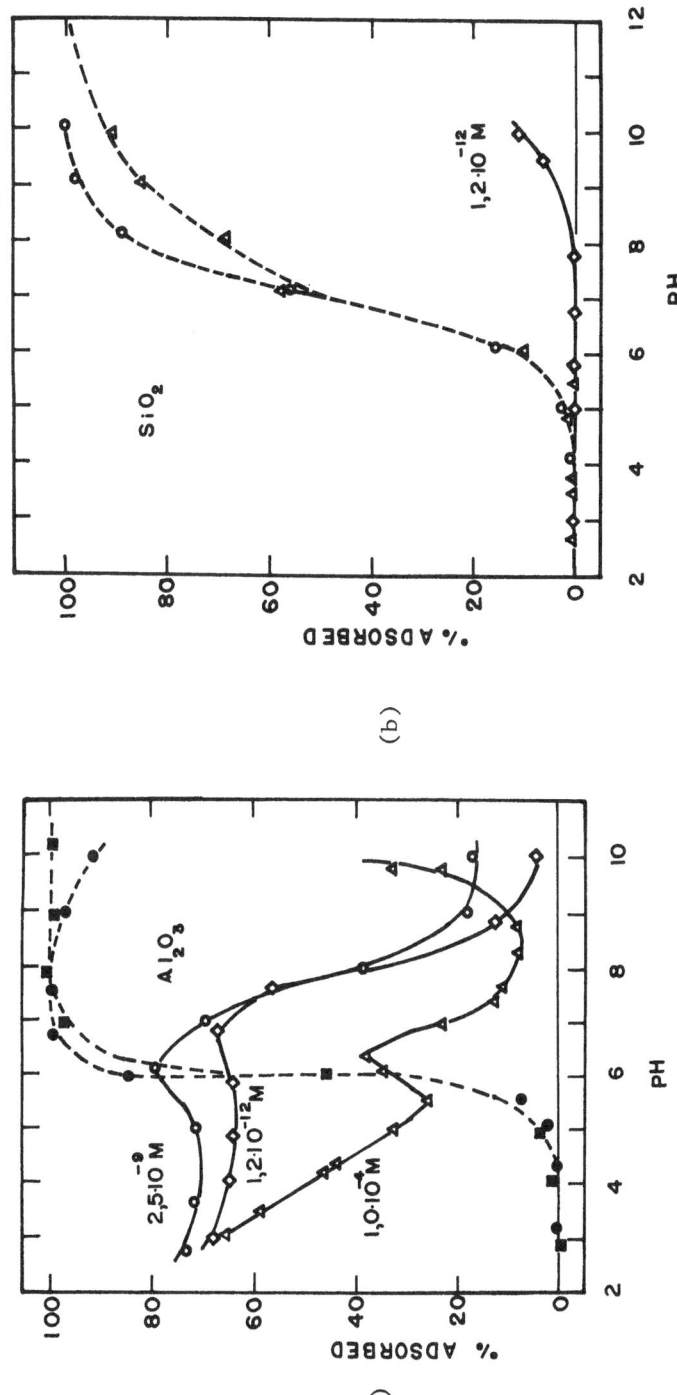

FIGURE 6: Comparing the adsorption characteristics of Co(II) and [Co(III)EDTA]⁻ complexes on hydrous solids, (a) γ-Al₂O₃ and (b) SiO2. In (a), both the open circles and the open diamonds were obtained by scintillation counting technique. The dashed lines are the same as those shown in Figure 2a. In (b), only the open diamonds were obtained by scintillation counting technique. The dashed lines are the same as presented in Figure 2b.

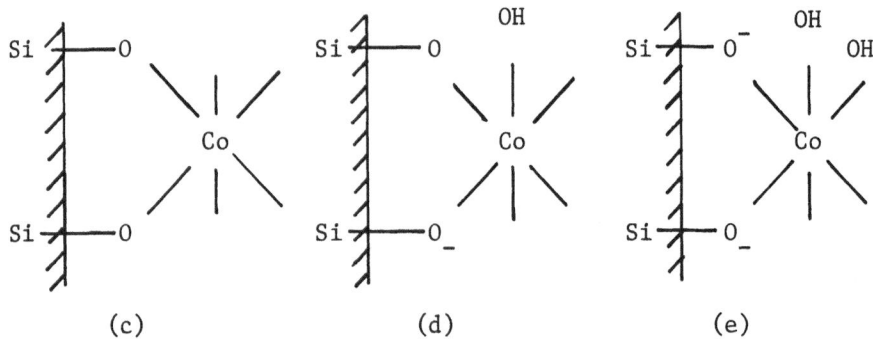

(c)   (d)   (e)

Therefore, the total amount of Co(II) adsorbed on an SiO_2
surface is described by the following expression

$$Co(II)_{ads} = \left[(\equiv SiO)_2Co\right] + \left[(\equiv SiO)_2CoOH^-\right] + \left[(\equiv SiO)_2Co(OH)_2^{2-}\right] \qquad (9)$$

Based upon the adsorption data presented in Figure 2b, the
values of free energy of adsorption calculated according to
Equation 4 vary from -2.0 Kcal/mol (pH 5) to -5.7 Kcal/mol (pH 10.0)
indicative of significant electrostatic effect. Electrostatic
effect at this level is unlikely since all adsorption data suggest
that electrostatic interaction is insignificant compared with
specific chemical adsorption. Moreover, there is reported evidence
suggesting the association of $CoOH^+$ and $Co(OH)_2$ species with
hydrous solids. James and Healy[3] suggested an interaction between
$CoOH^+$ and SiO_2 surface. Tewari showed from X-ray photoelectron
spectroscopic data[6] that the species adsorbed onto hydrous oxide
was $Co(OH)_2$.[6]

Based upon the results presented in Figures 1b and 2b, it is
possible to evaluate the equilibrium constants for the above
reactions. The calculated values vary from 5.8 (pH 5) to 7.8
(pH 8.2); 6.3 (pH 7.8) to 7.8 (pH 10); and 8.8 (pH 8.8) to
7.8 (pH 10), respectively for log β_2^s, log $\beta_{1,2}^s$ and log $\beta_{2,2}^s$,
respectively. These equilibrium constants can be translated into
free energy of adsorption, ΔG_{ads}. The average ΔG_{ads} values are -8.1,
-9.6 and -9.8 Kcal/mol, respectively, for surface complexes (c),
(d) and (e). James and Healy reported a specific chemical energy
of adsorption, ΔG_{chem}, for Co(II) on SiO_2 of -6.5 Kcal/mol
(Figure 2b).[3]

The $\equiv SiOH$ groups are probably unimportant in contributing to
Co(II) adsorption, since within the pH range of interest, the
fraction of $\equiv SiOH$ group decreases with increasing pH, whereas the
molar fraction of Co^{2+} species remains almost constant, at unity.
Therefore, it is not likely that the $\equiv SiOH$ group can directly

involve in the adsorption reaction, although Schindler et al.[22] had suggested a reaction between $\equiv SiOH$ and divalent cation, M^{2+}, as the major adsorption scheme.

One may note that there is $\equiv AlOCo^+$ but not $(\equiv AlO)_2Co$ surface complex formed in the Al_2O_3-Co(II) system, and $(\equiv SiO)_2Co$ but not $\equiv SiOCo^+$ surface complex formed in the SiO_2-Co(II) system. This can be partly attributed to the difference in surface acidity of the two hydrous solids. The total surface site densities are 2.1×10^{-4} mol/g (or 79 $\mathring{A}^2$/site) and 2.6×10^{-4} mol/g (or 208 $\mathring{A}^2$/site), respectively, for Al_2O_3 and SiO_2. By assuming that each site occupies a circular region, the distance between any two adjacent surface sites is 5 $\mathring{A}$ and 8 $\mathring{A}$, respectively, for Al_2O_3 and SiO_2. This difference in surface geometry, therefore, may discourage the linking of one Co^{2+} ion to two Al sites. However, based on presently available experimental results, it is not possible to unambiguously describe the mode of Co^{2+} ion attachment on hydrous solids.

Figure 5 illustrates the results of Co(II) adsorption on SiO_2 as predicted by the component reaction schemes. The calculated equilibrium constants of surface complexes and the values of free energy of adsorption for the corresponding surface complex formation of Co(II) on Al_2O_3 and SiO_2 are shown in Table 3.

Effect of Complex Formation

As shown in Figure 6, the adsorption characteristics of Co(II) differ significantly from those of [Co(III)EDTA]¯. The shape of the adsorption isotherm for Co(II) remains relatively unchanged regardless of the solid type, i.e. SiO_2 or Al_2O_3, although the major reaction schemes governing the adsorption process differ. The [Co(III)EDTA]¯ complexes are not adsorbed by SiO_2 to any significant extent at pH values less than 8.5. The adsorption isotherm of [Co(III)EDTA]¯ on Al_2O_3 differs from that of Co(II) on this solid. No Co(II) adsorption on Al_2O_3 was observed at pH < 5; whereas a significant amount of [Co(III)EDTA]¯ adsorbed on Al_2O_3 in this pH region.

The results clearly demonstrate that the nature of the chelate (or ligand) plays a rather important role in governing the adsorption characteristics of complexed metal ions, especially those with three-dimensional structure, such as [Co(III)EDTA]¯ complexes. For framework type complexes, the metal ion is located within the central position of the complex structure thereby rendering direct contact between the central metal ion and the hydrous surface impossible.

This finding is in accord with what was suggested by Davis and Leckie.[27] Ligands not adsorbed to oxide surfaces do not

Table 3. Values of Stability Constant and ΔG_{ads} of Surface Complexes

Equilibrium	K	Log K	ΔG_{ads} 25° C (Kcal/mol)
$Co^{2+} + \equiv AlOH \rightleftharpoons \equiv AlOCo^+ + H^+$	$*K_1^s$	-3.3	4.4
$Co^{2+} + 2(\equiv AlOH) \rightleftharpoons (\equiv AlO)_2Co^+ + 2H^+$	$*\beta_2^s$	-7.3	10.2
$Co^{2+} + 2(\equiv Si\bar{O}) \rightleftharpoons (\equiv SiO)_2Co$	β_2^s	6.1	-8.1
$CoOH^+ + 2(\equiv Si\bar{O}) \rightleftharpoons (\equiv SiO)_2CoOH$	$\beta_{2,1}^s$	7.2	-9.6
$Co(OH)_2 + 2(\equiv Si\bar{O}) \rightleftharpoons (\equiv SiO)_2Co(OH)_2$	$\beta_{2,2}^s$	7.4	-9.8
$CoY^- + \equiv AlOH_2^+ \rightleftharpoons (\equiv AlOH_2)CoY$	K_1^s	3.3	-4.4
$CoY(H_2O)^- + \equiv AlOH_2^+ \rightleftharpoons (\equiv AlOH_2)CoY(H_2O)$	K_2^s	4.4	-5.9
$CoY(H_2O)^- + \equiv AlOH \rightleftharpoons (\equiv AlOH)CoY(H_2O)^-$	K_3^s	3.4	-4.5
$CoY(OH)^{2-} + \equiv AlO^- \rightleftharpoons (\equiv AlO)CoY(OH)^{3-}$	K_4^s	3.6	-4.8

induce adsorption; therefore complex formation between non-adsorbable ligand and metals reduce the extent of metal adsorption. Vydra and coworkers[28] reported non-adsorption of anionic complexes formed from multivalent anionic ligands such as EDTA, and citrate on SiO_2 surfaces over a wide pH range. Dalang and Stumm[2] also reported that electrostatic interaction plays an important role in the adsorption of cationic cobalt complexes such as $Co(NH_3)_6^{3+}$, $CO(en)_3^{3+}$ and $Co(NH_3)_5Cl^{2+}$ on SiO_2. Forbes et al.[7] demonstrated that $[Co(NH_3)_6]^{3+}$ and $[Co(NH_3)_5(H_2O)]^{3+}$ were not adsorbed onto geothite due in part to the strong covalency of the $[Co(III)-OH]^{+2}$ bond that precludes the possible sharing of the hydroxo ligand with the solids.

It is interesting to note that the solid type plays a more important role in the adsorption of metal complexes than in the adsorption of uncomplexed metal ions. Elliott and Huang[15] have demonstrated that Cu(II)-Aspartate complexes, namely $[CuAsp]^\circ$ and $[Cu(Asp)_2]^{2-}$, were adsorbed by γ-Al_2O_3 but not by SiO_2 and TiO_2. However, it is known that metal ions could be adsorbed by most hydrous solids including SiO_2 and TiO_2.[1,11-12,29]

Adsorption of [Co(III)EDTA]⁻ Complexes

It is believed that, in a heterogeneous solution, hydrated [Co(III)EDTA] complexes such as $[Co(III)EDTA(H_2O)]^-$ and its corresponding conjugate base may exist. This is the cause of the difference in the extent of adsorption measured independently by spectrophotometry and the scintillation technique. By comparing the adsorption density of [Co(III)EDTA]⁻ on SiO_2 presented in Figures 4b and 6b, it becomes clear that the adsorption peaks observed in the pH range from 5 to 8.5 with spectrophotometry do not represent real adsorption reactions but rather are the result of reduction in the extinction coefficient (or absorbance) due to the presence of $[Co(III)EDTA(H_2O)]^-$ complex whose extinction coefficient at 540 nm is much lower than that of [Co(III)EDTA]⁻. Shimi and Higginson reported that $[Co(III)EDTA(H_2O)]^-$ is relatively unstable with respect to its unhydrated counterpart, i.e. [Co(III)EDTA]⁻.[17] However, the system studied by Shimi and Higginson was a homogeneous solution. The presence of solids such as SiO_2 and Al_2O_3 may catalyze the formation of hydrated $[Co(III)EDTA(H_2O)]^-$ complexes, although no information as such has yet been reported.

Adsorption experiments conducted with ^{60}Co isotopes are not affected by the optical properties of the Co(III) species and should account for all concentrations that disappear from the solution due to adsorption reactions. The results presented in Figures 4 and 6 demonstrate that [Co(III)EDTA]⁻ complex is adsorbed on both the Al_2O_3 and the SiO_2 surface at pH > 8. Since the

concentration of Co(III) used was at trace quantity, i.e.
1.2×10^{-12} M and there was an excess amount of EDTA present to
suppress the precipitation reaction, the removal of Co(III) from
the solution at pH > 8 is due to an adsorption reaction rather
than chemical precipitation.

In connection with the suggestion of Shimi and Higginson,[17]
that hydrated $[Co(III)EDTAH_2O]^-$ species are not stable, one may
tentatively assume $[Co(III)EDTA]^-$ as the predominant species
over the pH range studied, i.e. 3 to 10, and attempt to interpret
the adsorption results presented in Figure 4 or Figure 6.

Without distorting the rather symmetrical octahedral structure
of the $[Co(III)EDTA]^-$ complex, there are only three possible
distinct surface orientations for the attachment of $[Co(III)EDTA]$
on the solid surface as depicted in Figure 7:

1.   Orientation "a" wherein the three carboxyl oxygen atoms,
      O(I), coordinated with the central cobalt atom, and the
      three other oxygen atoms, O(II) of the carbonyl group are
      facing the solid surface (Figure 7a). Although the
      $[Co(III)EDTA]^-$ complex is a monovalent anion as a whole,
      within the structure of the complex ion, the formal charge
      of the nitrogen atoms is +1 and that of the cobalt atom
      is -3. Thus an arrangement on the surface would be facilitated
      when positive or neutral surface hydroxo groups, e.g. $\equiv AlOH_2^+$
      and $\equiv AlOH$, are available. The six oxygen atoms are almost
      coplanar, with the three carbonyl O(II) atoms being slightly
      closer to the surface than the three carboxyl O(I) atoms.
      This provides the most "mechanically" stable state and
      opportunity to form specific hydrogen bondings. The following
      modes of surface association are possible:

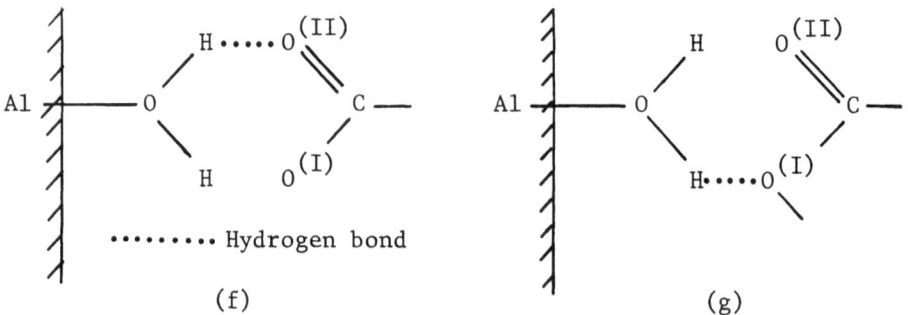

Due to the fact that carboxyl O(I) and carbonyl O(II) atoms are
not exactly coplanar, there is only one hydrogen bonding possible per
surface site, as shown. Similarly, for the neutral hyroxo group,
there is only one hydrogen bonding possible per surface site:

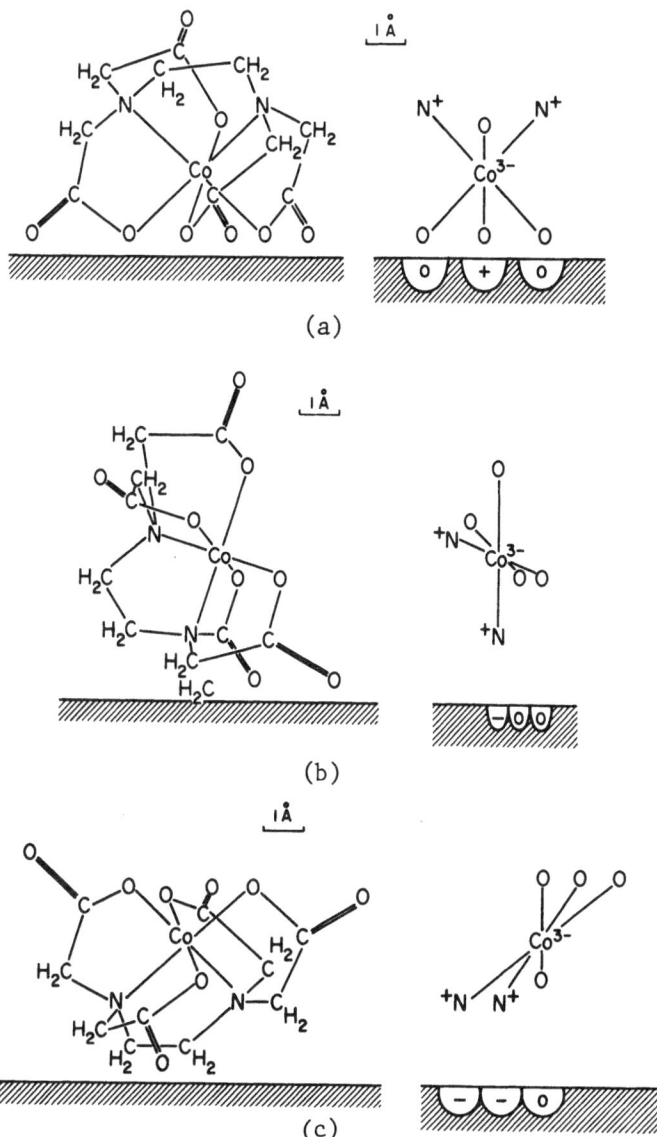

FIGURE 7: Schematic diagram showing the possible orientations of
[Co(III)EDTA]⁻ complex on hydrous oxide surface. (a)
most stable configurations with three O(I) atoms and
three O(II) atoms facing the solid; (b) meta-stable
configuration with one N atom and two O(I) atoms
facing the surface; (c) meta-stable configuration with
two N atoms and one O(I) atom facing the surface. The
symbols 0, + and - represent a neutral, positive and
negative surface, respectively.

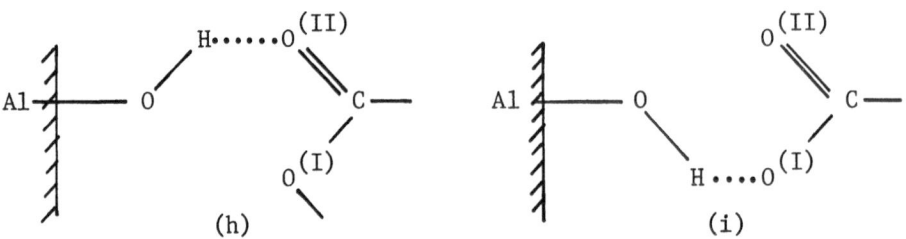

2.   Orientation "b" by which, conceptually, the complexed ion is
     located with its two N atoms and one carboxyl O(I) atom
     towards the surface (Figure 7b). Although the two CH_2
     groups stay mostly close to the solid surface, due to the
     electronegativities of carbon and hydrogen atoms (2.5 and
     2.1 respectively,) hydrogen bonding is unlikely between the
     CH_2 group and the surface hydroxo groups. Electrostatically,
     this orientation is most favored by a surface containing
     both neutral and negative hydroxo groups. Hydrogen bonds
     may be formed between the O(I) atom and the surface hydroxo
     group; however, "mechanically" this orientation is not
     stable. Furthermore, reaction between $\equiv AlOH$ and $[Co(III)EDTA]^-$
     does not seem to contribute to the adsorption process;
     otherwise, $[Co(III)EDTA]^-$ would be expected to sorb on the
     SiO_2 surface at the pH region when $\equiv SiOH$ is available, i.e.
     pH < 8. No adsorption of $[Co(III)EDTA]^-$ complex was found
     at pH 8.

3.   In orientation (c) the complex ion lies on the surface with
     two carboxyl O(I) atoms, and one nitrogen atom as touching
     point. Because the two N atoms have +1 formal charge, this
     orientation is most favored by a surface with both neutral
     and negative hydroxo groups. The opportunity to form specific
     chemical bondings is only via the carboxyl O(I) atom, since
     the CH_2 group, although being the closest of all to the
     surface, cannot form hydrogen bonding to secure the complex
     ion on the surface. Furthermore, "mechanically" this orienta-
     tion is not as stable as orientation "a".

     Based upon the above analysis, it appears that orientation
(a) contributes significant surface attachment. Orientations (b)
and (c) are not likely to result in the adsorption of $[Co(III)EDTA]^-$
complex to any significant degree. The results obtained with
^{60}Co measurement showed that $[Co(III)EDTA]^-$ ion is not adsorbed
on SiO_2 even when both the $\equiv SiOH$ and the $\equiv SiO^-$ hydroxo groups are
present, indicating that orientations (b) and (c) do not bring
about a stable surface complex. However, within the pH range
where both the $\equiv AlOH$ and the $\equiv AlO^-$ hydroxo groups are present,
$[Co(III)EDTA]^-$ complex is markedly adsorbed.

     Based upon the above discussion, the following reaction
schemes are suggested to describe the adsorption behavior of
$[Co(III)EDTA]^-$ complex:

$$CoY^- + \equiv AlOH_2^+ \rightleftharpoons (\equiv AlOH_2)CoY; \ K_1^s \qquad (9)$$

This reaction leads to the formation of surface complexes (f)
and/or (g). Reaction between CoY^- and $\equiv AlOH$ with the formation
of surface complexes (h) and/or (i) is not likely to occur to any
significant extent; otherwise CoY^- should be found significantly
adsorbed on SiO_2 surface at pH < 8.5 when $\equiv SiOH$ groups are present.
However equation 9 alone can not account for the total amount of
[Co(III)EDTA]⁻ adsorbed, since it will predict a decrease in the
extent of [Co(III)EDTA]⁻ adsorbed as pH increases. This suggests
that other forms of Co(III) complexes must be responsible for the
adsorption behavior observed.

In order to account for the adsorption density on Al_2O_3 in
the pH range from 5 to 8 (or adsorption region II) the following
reactions that involve [Co(III)EDTA(H₂0)]⁻ complex are suggested:

$$\equiv AlOH_2^+ + CoY(H_2O)^- \rightleftharpoons \left[(\equiv AlOH_2)(CoY(H_2O))\right]; \ K_2^s \qquad (10)$$

$$\equiv AlOH + CoY(H_2O)^- \rightleftharpoons \left[(\equiv AlOH)CoY(H_2O)\right]^-; \ K_3^s \qquad (11)$$

The following surface complexes are formed through the formation
of specific chemical bondings between the non-coordinated COO^-
group and the surface hydroxo group:

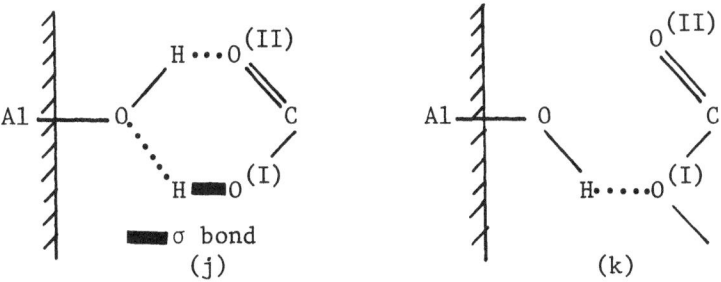

Surface complex (j) has one σ-bond and two hydrogen bondings;
whereas, (k) has only one hydrogen bonding per adsorption site. It
is therefore expected that ΔG_{ads} will be greater with surface complex
(j) than with (k) or others, i.e. (f to i). The calculated ΔG_{ads}
values for various reaction schemes are listed in Table 3. The
results give a log K_2^s value of 4.4 (or ΔG_{ads} = -5.9 Kcal/mol) as
opposed to a log K_3^s of 3.4 (or ΔG_{ads} = -4.5 Kcal/mol) and log K_1^s of
3.3 (or ΔG_{ads} = -4.4 Kcal/mol).

It is further noted that neutral surface hydroxo groups,
i.e. $\equiv AlOH$ react with [Co(III)EDTA(H₂0)]⁻ complex ions but not

with [Co(III)EDTA]$^-$. This is expected since of the three orientations possible for attaining a stable arrangement on hydrous
oxide surface, only orientation (a) is stable. As discussed
above, orientation (a) prefers positive surface sites to neutral
ones due to electrostatic effect. In adsorption region II, i.e.
5.5 < pH < 7.5, the amount of AlOH$^+$ decreases, with increasing pH
thereby discouraging the formation of surface complexes (f)
and/or (g). However, once hydrated, one of the two O(I) oxygen
atoms, located opposite to the N-atom, becomes uncoordinated with
the central metal atom, i.e. Co(III). In order to maintain the
coordination state of the cobalt atom, an H$_2$O molecule is brought
in to substitute for the uncoordinated -COO$^-$ group. The adsorption
behavior of [Co(III)EDTA(H$_2$O)]$^-$ depends on the linking mechanism
now available from the uncoordinated -COO$^-$ group. The total
change of free energy of adsorption, ΔG_{ads} is greater for the
uncoordinated group than for the coordinated one.

However, these above reaction schemes still cannot totally
describe the total adsorption density observed with γ-Al$_2$O$_3$ and
SiO$_2$ particularly at pH > 8.5. In order to account for increasing
adsorption density at pH > 8.5 for both solids studied, the
following reactions are therefore suggested:

$$\equiv AlO^- + \left[CoY(OH)\right]^{2-} \rightleftharpoons \left[(\equiv AlO)CoY(OH)\right]^{3-} ; \ K_4^s \qquad (12)$$

Equation 12 suggests the formation of a surface complex of the
following form:

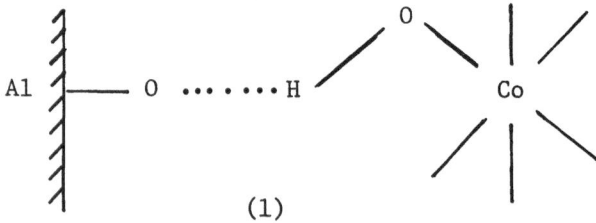

(1)

Based upon the data presented in Figure 4a or 6a, the equilibrium constants and their corresponding values of change of free
energy of adsorption, ΔG_{ads} are calculated. The results of this
analysis are presented in Table 3. Figure 8 shows the calculated
isotherms based upon the four equilibrium reactions for γ-Al$_2$O$_3$
only. The agreement between calculated and observed adsorption
data is excellent.

The speciation of [Co(III)EDTA] complexes (at 10^{-4} M), was
based on the data presented in Figure 4b. Since both [Co(III)EDTA]$^-$
and [Co(III)EDTA(H$_2$O)]$^-$ are not adsorbed by hydrous SiO$_2$ while

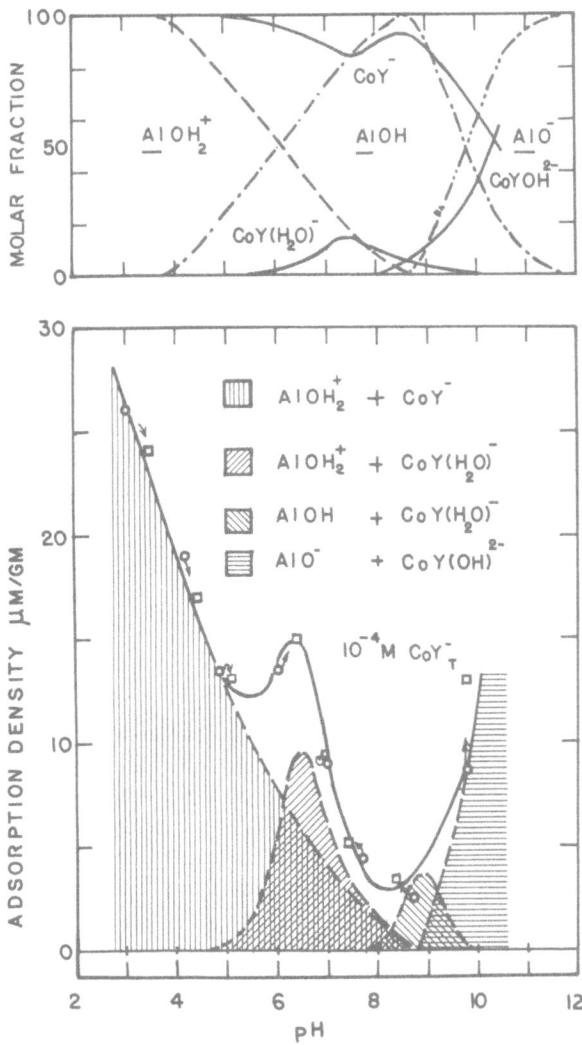

FIGURE 8: Formation of surface complexes in [Co(III)EDTA]⁻ Al₂O₃
system. The speciation of [Co(III)EDTA] species
was based on results presented in Figure 4b. The
adsorption density contributed from each reaction
scheme adds up to the total adsorption density observed.
The open circles and the open squares stand for
adsorption density measured from the same sample but
at two different reaction time intervals. The open
circles were overnight, whereas the open squares were
two day old samples. The arrows connect the data
points obtained from the same sample.

$[Co(III)EDTA(OH)]^{2-}$ is adsorbable on a negatively charged SiO_2 surface, the difference in concentrations between the original and the residual during the adsorption experiments observed for the SiO_2-$[Co(III)EDTA]$ system can be attributed to the reduction of adsorbance due to the formation of $[Co(III)EDTA(H_2O)]^-$ complex at pH < 8.5 and $[Co(III)EDTA(OH)]^{2-}$ complex at pH > 8.5. It is noted that at pH 8.5, both hydrated $[Co(III)EDTA]$ complexes are present at equimolar amounts. This means that the acidity constant, $pK_{a2'}$ for $[Co(III)EDTA(H_2O)]^-$ deprotonation equilibrium has a value of 8.5, which is in excellent agreement with what was reported by Shimi and Higginson.[17] This finding serves to enhance the speculation that in heterogeneous solutions such as those two studied, the solid present may catalyze the formation of hydrated $[Co(III)EDTA]$ species, i.e. $[Co(III)EDTA(H_2O)]^-$ and $[Co(III)EDTA(OH)]^{2-}$. A further analysis of these hydrated species by more advanced and direct techniques such as NMR is deemed essential to verify the existence of $[Co(III)EDTA(H_2O)]^-$ and $[Co(III)EDTA(OH)]^{2-}$ in a heterogeneous solution, although all evidence observed in this study does suggest the presence of hydrated $[Co(III)EDTA]$ complexes.

CONCLUSION

    Based upon the results obtained in this study, it is apparent that specific chemical interaction through the formation of hydrogen bonding and σ-bonding plays an important part in the adsorption of Co(II) and $[Co(III)EDTA]^-$ complexes. This is indicated by the fact that the order of magnitude of adsorption energy, ΔG_{ads}, is significantly larger than that of the electrostatic one.

    Although the shape of the adsorption isotherm, i.e. adsorption density versus pH of Co(II) in particular and other hydrolyzable cations, M(II) in general, on hydrous solids may be similar to each other except for the difference in pH of abrupt adsorption, the reaction mechanism could be entirely different. This is demonstrated by the Co(II) adsorption on SiO_2 and Al_2O_3, respectively. For Co(II) adsorption on Al_2O_3, only a 1:1 surface complex of the type $\equiv AlOCo^+$ is formed according to the following reaction:

$$\equiv AlOH + Co^{2+} \rightleftharpoons \equiv AlOCo^+ + H^+; \quad *K_1^s$$

whereas for Co(II) adsorption on SiO_2, three different surface complexes are possible, i.e. $(\equiv SiO)_2Co$, $(\equiv SiO)_2CoOH^{-1}$ and $(\equiv SiO)_2Co(OH)_2^{2-}$ according to the reactions:

$$2\equiv SiO^- + Co^{2+} = (\equiv SiO)_2Co; \quad \beta_{2,0}^s$$

$$2{\equiv}SiO^- + CoOH^+ = ({\equiv}SiO)_2CoOH^{-1}; \beta^s_{2,1}$$

$$2{\equiv}SiO^- + Co(OH)_2 = ({\equiv}SiO)_2Co(OH)_2^{2-}; \beta^s_{2,2}$$

The adsorption behavior of a strong framework-like metal complex such as [Co(III)EDTA]⁻ onto hydrous solids differs significantly from that of its uncomplexed counterpart. Although the extent of metal complex adsorption is substantially affected by the adsorbability of the free ligand, the adsorption behavior of the complexed metal does not exactly copy that of the free ligand.

It is seen that EDTA is not very adsorbable on SiO_2 nor are [Co(III)EDTA]⁻ complexes. However, [Co(III)EDTA]⁻ complexes, presumably [Co(III)EDTA(OH)]²⁻, become significantly adsorbed on SiO_2 at pH > 8.5. For Al_2O_3, three adsorption regions, governed by different adsorption schemes were observed. The adsorption behavior of free EDTA species on Al_2O_3 is typical of anionic ligands.[12]

Based upon stereochemical consideration in addition to the possibility of specific chemical and electrostatic interactions, four reaction schemes are suggested to describe the adsorption behavior of [Co(III)EDTA]⁻ complexes on γ-Al_2O_3:

$$CoY^- + {\equiv}AlOH_2^+ \rightleftharpoons ({\equiv}AlOH_2)CoY; K^s_1$$

$$CoY(H_2O)^- + {\equiv}AlOH_2^+ \rightleftharpoons \left[({\equiv}AlOH_2)CoYH_2O\right]; K^s_2$$

$$CoY(H_2O)^- + {\equiv}AlOH \rightleftharpoons \left[({\equiv}AlOH)CoY(H_2O)\right]^-; K^s_3$$

$$CoY(OH)^{2-} + {\equiv}AlO^- \rightleftharpoons ({\equiv}AlO)CoY(OH)^{3-}; K^s_4$$

It is further proposed that in a heterogeneous solution, the presence of solids such as Al_2O_3 and SiO_2 may enhance the stability of hydrated [Co(III)EDTA] species, i.e. [Co(III)EDTA(H_2O)]⁻ and [Co(III)EDTA(OH)]²⁻. This finding was based upon adsorption experiments using both spectrophotometry and a scintillation technique.

ACKNOWLEDGMENT

This research work was supported by a grant, ENG 77-27379, from the Water Resources, Urban and Environmental Engineering Program of the National Science Foundation.

REFERENCES

1. H. Hohl and W. Stumm, J. Colloid Interface Sci. 55:281 (1976).
2. F. Dalung and W. Stumm, J. Colloid Interface Sci.
    M. Kerker, R.L. Rowell and A.C. Zettelmoyer, eds., Academic
    Press, Vol. IV, 157 (1976).
3. R.O. James and T.W. Healy, J. Colloid Interface Sci. 40:42
    (1972).
4. J.E. Duval and M.H. Kurbatov, J. Phys. Chem. 56:982 (1952).
5. R.O. James, P.J. Stiglich and T.W. Healy, Disc. Faraday Soc.
    (1975).
6. P.H. Tewari and W. Lee, J. Colloid Interface Sci. 52:77 (1975).
7. E.A. Forbes, J. Colloid Interface Sci. 49:403 (1974).
8. D.L. Dugger, J.H. Stanton, B.N. Irby, B.L. McConnell,
    W.W. Cummings and R.W. Maatman, J. Phys. Chem. 68:757 (1964).
9. P. Loganathan and R. Burau, Geochim. et Cosmo Chim. Acta
    37:1277 (1973).
10. M. MacNaughton and R. James, J. Colloid Interface Sci. 47:431
    (1968).
11. J. Vuceta, Ph.D. Thesis, California Institute of Technology,
    1976.
12. M.M. Benjamin, Ph.D. Thesis, Stanford University, 1979.
13. H.A. Elliott and C.P. Huang, Environment International 2:145
    (1979).
14. H.A. Elliott and C.P. Huang, J. Colloid Interface Sci. 70:29
    (1979).
15. H.A. Elliott and C.P. Huang, J. Envir. Sci. Tech. 14(1):87
    (1980).
16. L.E. Orgel, "An Introduction to Transition Metal Chemistry
    Ligand Field Theory," John Wiley & Sons, Inc., New York
    (1980).
17. I.A.W. Shimi and W.C.E. Higginson, J. Amer. Chem. Soc. 80:260
    (1958).
18. L.G. Sillens and A.E. Maxwell, "Critical Stability", Vol. 3,
    Plenum Press, New York (1977).
19. C.P. Huang and W. Stumm, J. Colloid Interface Sci. 43:409
    (1973).
20. W. Stumm, R. Kummert and L. Sigg, Croat. Chim. Acta. 52:120
    (1980).
21. P.W. Schindler and H.R. Kamber, Helu Chim. Acta. 51:1781
    (1968).
22. P.W. Schindler, B. Furst, R. Dick and P.U. Wolf, J. Colloid
    Interface Sci. 55:469 (1976).
23. G.H. Bolt, J. Phys. Chem. 61:1166 (1957).
24. C.P. Huang, in: "Adsorption at Solid-Liquid Interface,"
    M. Anderson and A. Rubin, eds., Am. Arbor Sci. (1981).
25. K.L.E. Kaiser, Water Res. 7:1465 (1973).
26. J. Rubio and E. Matijevic, J. Colloid Interface Sci. 68:408
    (1979).

27. J.A. Davis and J.O. Leckie, Environ. Sci. Tech. 12:1309
     (1978).
28. F. Vydra, Ralanta 10:753 (1963).
29. C.P. Huang, H.A. Elliott and R.M. Ashmead, J. Water Pollution
     Control Fed. 49:745 (1977).

THE ADSORPTION OF ALKALINE-EARTH METAL IONS

AT THE RUTILE/AQUEOUS SOLUTION INTERFACE

D.W. Fuerstenau, D. Manmohan and S. Raghavan

Department of Materials Science
and Mineral Engineering
University of California
Berkeley, California 94720

ABSTRACT

The adsorption characteristics of alkaline-earth metal ions at the rutile/aqueous solution interface have been investigated, utilizing electrokinetic and direct adsorption techniques. The results of the electrokinetic investigations indicate that the alkaline-earth cations behave as indifferent ions at low concentration (less than 10^{-5} M) but exhibit strong specific adsorption at higher concentrations, resulting in the reversal of charge. The specific adsorption of these inorganic cations is particularly significant at pH values greater than the point of the zero charge of rutile. The affinity of these ions for the rutile surface has been found to follow the sequence:

Ba > Sr > Ca > Mg

This order correlates with the ionic radii and hence the ease with which the ions dehydrate. The results obtained from the adsorption experiments are in close agreement with the results of electrokinetic measurements.

INTRODUCTION

Oxides are abundantly available and have always been an important source of raw materials for both minerals and metals. However, it is often difficult to separate oxides selectively by flotation because such separations are generally based on differences in the surface charge of the various oxides,[1] unless chemisorbing flotation collectors (surfactants) can be found.[2]

Often, multivalent ions are present in flotation slurries and
these ions can either facilitate or complicate surface reactions
in flotation, mainly through their role as activators in causing
collectors to adsorb at mineral surfaces.[3] Similarly the behavior
of aqueous oxide suspensions is of technological importance in
ceramic processing and in the control of pigment systems. Accord-
ingly, the present investigation was undertaken to better under-
stand the surface properties of oxides in the presence of the
inorganic alkaline-earth ions, and particularly to study the
nature of specific adsorption of these ions.

The solid oxide used for this research was rutile, one of
the polymorphic modifications of titania (TiO_2). The reasons for
choosing titania as the model oxide for study are many. It
exhibits a reversal of charge at about neutral pH--which provides
a good pH range within which to work, and also enables one to
investigate adsorption behavior on both a positive and a negative
surface. This material also has an extremely low solubility in
water and can be considered to be insoluble. It is readily
available and has technological importance, especially in the
paint pigment industry. Finally, there is also a large amount of
published information available on titania and its surface properties.[4-11]

To systematically study the adsorption behavior of alkaline-
earth ions and the effect that they have on the surface charge
characteristics of titania, the electrokinetic behavior of rutile
was investigated in the presence of the nitrate salts of barium,
strontium, calcium and magnesium. Alkaline-earth metal ions were
chosen for this study not only because of their technological
importance but also because of their divalent nature and, consequently, their pronounced tendency to adsorb. Importantly, these
ions do not hydrolyze appreciably in the pH range 4 - 10. Further,
since these cations belong to the same group of the periodic
table, it is possible to study systematically how ionic size and
hydration phenomena affect adsorption. Our main experimental
method involved determination of how these electrolytes affected
the electrokinetic potential of rutile. Electrokinetic experiments
have long been known to provide a clear demonstration of the
specific adsorption of nonhydrolyzing multivalent ions on oxides
from aqueous solutions.[12,49] With streaming potential measurements, Purcell and Sun[5] demonstrated that sulfate ions and calcium
ions can shift the pH of the reversal of the zeta potential of
rutile to pH values lower and higher than the point-of-zero
charge, respectively. Although the major part of this work is
concerned with the electrokinetic behavior of rutile, a few
direct adsorption measurements were carried out to corroborate
the findings.

BACKGROUND

The surface properties of rutile have been extensively
studied by colloid and surface chemists, including the determination of the point-of-zero charge (PZC) and the surface charge by
electrophoretic and titrimetric methods, the measurement of the
adsorption of various inorganic and organic ions at the rutile
solution interface, and the study of the energetics of the titania/
aqueous solution interface by heat of immersion measurements.

A recent review of the surface chemical properties of rutile
indicates that the PZC of rutile samples occurs around a pH of
5.5 and that the PZC is markedly affected by adsorbed impurities
such as chloride and sulfate which are inevitably present because
of the mode of preparation.[12]   Berubé and de Bruyn[8,9] obtained
evidence for the specific adsorption of alkali cations, on rutile
from indirect capacity calculations. The specific adsorption of
these cations seems to decrease in the order: Li > Na > Cs.
There is evidence in the literature[8] that inorganic ions Cl^-,
NO_3^- and ClO_4^- behave similarly and indifferently at the rutile-
aqueous solution interface. Similar to all other systems involving hydrolyzing cations, James and Healy[6] found isoelectric
points of TiO_2 to occur at more than one pH value in the presence
of Co(II), La(III) and Th(IV) salts and explained this in terms
of the specific adsorption of multivalent cations on TiO_2 followed
by the nucleation and precipitation of metallic hydroxides in the
interfacial region. Wiese and Healy[13] working with the TiO_2/
$Al(NO_3)_3$ system found Al(III) ions to exert a similar expected
influence on the surface properties of TiO_2. However, conditions
used in the present investigation were such that ion hydrolysis
should not be involved in the adsorption process.

Reported studies on the interaction of alkaline-earth cations
with rutile are virtually nil. An investigation by Malati and
Smith[47] indicates that the affinity sequence for these cations,
as measured through adsorption measurements, follows the order Ca
> Sr > Ba. Since the affinity order that we have observed for
rutile was Ba > Sr > Ca > Mg, the opposite of that found by
Malati and Smith, examination of the reported affinity sequences
in various systems appears to be useful.

There are conflicting reports in the literature as to the
affinity sequence of these cations for Fe_2O_3 surfaces. Breeuwsma
and Lyklema[14] conducted acid-base titrations on Fe_2O_3 samples
(PZC = 8.5) in the presence of the nitrate salts of alkaline-
earth ions. Their results indicate that at high concentrations
of these salts (5 x 10^{-2} M), the ions exhibit the order Mg > Ca >
Sr = Ba. In their words, "the strikingly high surface charge in
the presence of Mg^{2+} is obviously a _radius effect_ and not a

charge effect". They have speculated that the material exhibits a "chemical" kind of porosity (which defies detection by gas adsorption studies) and due to the fact that the crystal ionic radii of Mg^{2+} (0.65 Å) and Fe^{3+} (0.64 Å), are almost identical, Mg^{2+} can penetrate the surface much more readily. This explanation assumes that the ions adsorb in a completely dehydrated form. At lower ionic strength, they observed that the affinity of all the alkaline-earth ions is about the same. In spite of this high adsorbability as exhibited by charge measurements, Breeuwsma found Mg^{2+} ion to be only just as effective as Ba^{2+} in coagulating Fe_2O_3 sols. On the other hand, Kinneburgh[15] observed that freshly precipitated iron oxide gel (PZC = 8.1) exhibits the following selectivity sequence: Ba > Ca > Mg.

In an investigation of the adsorption of the alkali-earth cations on quartz at a pH of 10.5 using a radiotracer technique, Malati and Estafan[16] found the adsorbability to decrease in the order Ba > Sr > Ca, which is the reverse order for the hydrated radii of these ions. This led them to conclude that in the electrical double layer at the quartz/solution interface, the adsorbed alkaline-earth ions are hydrated. Tadros and Lyklema[17] measured the surface charge density of silica in the presence of the chlorides of Mg, Ca, Sr and Ba and concluded from their experimental results that the alkaline-earth metal ions exhibit specific adsorption in SiO_2 above a solution pH of 7.5. The extent of adsorption of these cations was found to follow the order: Ba > Ca > Sr > Mg.

Electrophoretic and adsorption measurements with δ-MnO_2 and γ-MnO_2 substrates have shown that the selectivity sequence of alkaline-earth cations on these oxides follows the order Ba > Sr > Ca > Mg.

The adsorbability of alkaline-earth metal cations on the alumina surface seems to follow the order Mg > Ca > Sr > Ba, which is just the reverse of the selectivity sequence in SiO_2 and MnO_2 surfaces. An explanation for the increased adsorbability of the most hydrated ion (Mg^{2+}) compared to the least hydrated ion (Ba^{2+}) has been offered in terms of the water structure at interfaces and the structure-making or breaking properties of the ions.[18]

Malati and Smith[47] recently reported on the adsorption of alkaline-earth ions on rutile and anatase. They observed the adsorption sequence to be Ca > Sr > Ba in both cases. However, their anatase analyzed 5.9% SO_3 and their rutile contained 1-2% chlorine, indicating quite impure starting materials. In addition, the adsorption was carried out with the solids being suspended in 0.1 M KNO_3. Thus, the adsorption order may reflect either the behavior of impure adsorbents or the effects of a high surface charge due to the high ionic strength.

Table 1. Order of Adsorption of Alkaline-Earth Cations on Solids from Aqueous Solution

Material	Adsorption Order	Experimental Method	Reference
Gold	Ba > Ca	coagulation value	Freundlich and von Elissatoff (22)
AgI	Ba > Sr > Ca > Mg	coagulation value	Kruyt and Klompe (23)
As₂S₃	Ba > Ca > Sr > Mg	coagulation value	Freundlich (24)
Cation Exchange Resin	Ba > Sr > Ca > Mg	ion exchange	Kunin (25)
Al₂O₃	Mg > Ca > Sr > Ba	potentiometric titration	Huang and Stumm (18)
Al₂O₃	Mg > Ca > Sr > Ba	adsorption	Kinniburgh (15)
Fe₂O₃	Mg > Ca > Sr = Ba	potentiometric titration (high ionic strength)	Breeuwsma and Lyklema (14)
Fe₂O₃	Mg = Ca = Sr = Ba	potentiometric titration (low ionic strength)	Breeuwsma (26)
Fe₂O₃	Ba > Ca = Mg	adsorption	Kinniburgh (15)
TiO₂ (Rutile)	Ca > Sr > Ba	adsorption	Malati and Smith (47)
TiO₂ (Rutile)	Ba > Sr > Ca > Mg	electrophoresis	This work
TiO₂ (Rutile)	Ba > Sr > Ca > Mg	adsorption	This work
TiO₂ (Anatase)	Ca > Sr > Ba	adsorption	Malati and Smith (47)
γ-MnO₂	Ba > Sr > Ca	adsorption	Gabano et al. (27)
δ-MnO₂	Ba > Sr > Ca	adsorption	Malati and Gray (28)
Hydrous MnO₂	Ba > Sr > Ca > Mg	adsorption	Murray (29)
SiO₂ (Silica)	Ba > Sr > Ca	adsorption	Malati et al. (30)
SiO₂ (Silica)	Ba > Ca > Sr > Mg	potentiometric titration	Tadros and Lyklema (17)
SiO₂ (Quartz)	Ba > Sr > Ca	adsorption	Malati and Estafan (16)
SiO₂ (Quartz)	Ba > Sr > Ca	heterocoagulation	von Buzagh (31)
Zirconium phosphate	Ba > Sr > Ca > Mg	distribution coefficient	Amphlett et al. (32)
Clinoptilolite	Ba > Sr > Ca > Mg	ion exchange	Ames (33)
Beidelite	Ba > Sr > Ca > Mg	ion exchange	Marshall (34)

Table 2.   Order of Adsorption of Alkali-Metal Cations on Solids from Aqueous Solution

Material	Adsorption Order	Experimental Method	Reference
Gold	K > Na	coagulation value	Freundlich and von Elissatoff (22)
Sulfur	Cs > Rb > K > Na > Li	coagulation value	Freundlich (35)
AgI	Rb > K > Na > Li	coagulation value	Kruyt and Klompe (23)
AgI	Rb > K > Na > Li	coagulation value	Lyklema (36)
As$_2$S$_3$	K > Na > Li	coagulation value	Freundlich (24)
Fe$_2$O$_3$	Li > Na = K	potentiometric titration	Breeuwsma (26)
Fe$_2$O$_3$	Li > Na > K > Cs	coagulation value	Dumont and Watillon (37)
TiO$_2$ (Rutile)	Li > Na > Cs	potentiometric titration	Berube and de Bruyn (8)
Rutile (high energy)	Li > Na > K > Cs > Rb	adsorption	Dumont and Watillon (48)
Rutile (low energy)	Rb > Cs > K > Na > Li	adsorption	Dumont and Watillon (48)
MnO(OH)$_2$	Cs > K > Na > Li	ion exchange	Gerevini and Somigliana (38)
δ-MnO$_2$ (Manganite IV)	Cs > Na	adsorption	Stumm, Huang, Jenkins (39)
β-MnO$_2$ (Pyrolusite)	Na > Cs	adsorption	Stumm, Huang, Jenkins (39)
δ-MnO$_2$	Na > K	adsorption	Murray (29)
SiO$_2$ (Quartz)	K > Na > Li	heterocoagulation	von Buzagh (31)
Sulfonic Acid Resin	Cs > Rb > K > Na > Li	ion exchange	Boyd, Schubert, Adamson (40)
Cation Exchange Resin	Cs > Rb > K > Na > Li	ion exchange	Kunin (25)
Zirconium Phosphate	Cs > K > Na > Li	perm selectivity	Alberti (41)
Zirconium Phosphate	Cs > Rb > K > Na > Li	distribution coefficient	Amphlett et al. (32)
Clinoptilolite	Rb > K > Na > Li	ion exchange	Ames (33)
Permutit	Cs > Rb > K > Na > Li	ion exchange	H. Jenny (44)
Beidelite	Cs > Rb > K > Na > Li	ion exchange	Marshall (34)
Montmorillonite	Cs > Rb > K > Na > Li	potentiometric titration	Schachtschabel (42)
Attapulgite	K > Na > Li	ion exchange	Merriam and Thomas (43)
Zeolite	Cs > Rb > K > Na > Li	ion exchange	Jenny (44)
Ultramarine	Na > Li > Rb > Cs	ion exchange	Barrer (45,46)
Analcite	K > Na > Rb > Li > Cs	ion exchange	Barrer (45,46)
Mordenite	Na > K > Li	ion exchange	Barrer (45,46)
Chabazite	K > Na > Cs > Li	ion exchange	Barrer (45,46)
Faujasite	Cs > Rb > K > Na	ion exchange	Barrer (45,46)

Table 1 summarizes the results of various researchers on the relative affinity of alkaline-earth ions for various adsorbents. As can be seen from Table 1, generally the adsorption order seems to be Ba > Sr > Ca > Mg but still there are conflicting findings. Clearly, the interaction of alkaline-earth ions with oxides (and other materials) has been the subject of many investigations but the mechanism of the process is yet to be established to any degree of certainty.

The case for the adsorption order of the alkali metal cations shows similar conflicting and interesting patterns, as can be seen by the summary (Table 2) of various observations on the affinity series of alkali ions for various adsorbents. For example, Stumm et al. demonstrated the reversal in adsorption order on δ-MnO$_2$ (manganite IV) and β-MnO$_2$ (pyrolusite): for δ-MnO$_2$, Cs$^+$ > Na$^+$ but for β-MnO$_2$, Na$^+$ > Cs$^+$. They interpreted this in terms of the effect of the electrostatic field strength on the structure of the adjacent water layer at the solid surface and its role in cation adsorption.

Recently Dumont and Watillon[48] prepared two different types of rutile surfaces, one having a low energy and the other a high energy surface, and the lyotropic sequences are opposite on the two adsorbents. On the high energy surface Li$^+$ adsorbs more strongly than Rb$^+$, and for the low energy surface, the reverse was observed. Considerable work remains to be done before these phenomena are unambiguously understood.

EXPERIMENTAL MATERIALS AND METHODS

The sample of rutile (designated as Lot CLDD/924) used for the present study was supplied by Tioxide International through the courtesy of Dr. G.D. Parfitt. This synthetic sample had been prepared by the hydrolysis of redistilled TiCl$_4$, followed by heating in air at 450°C for two hours. The pH of the aqueous dispersion of the as-received sample was about 3, which indicated a strong retention of hydrochloric acid on the surface after preparation.

Preparation of this material for interfacial studies was done in the following manner. The excess acid was first neutralized by sodium hydroxide to bring the pH to about 7. The suspension was then vigorously stirred in triply distilled water for an hour and left overnight, after which the supernatant was decanted. This procedure was repeated about thirty times until the atomic absorption spectrophotometer showed no sodium present. After the washed rutile had been dried at 100°C, it was characterized by spectrographic analysis (for trace metal impurities), X-ray powder diffraction analysis, and B.E.T. surface area measurement.

Table 3.   Impurity Elements, Reported as
           Oxides, in the Synthetic Rutile,
           Lot CLDD/924

Element	Weight %
Si	0.015
Mg	0.001
Al	0.004
Cu	0.001
Ca	0.007
Ba	0.001
S	0.03

NOTE
Analyzed by the American Spectrographic Co., San Francisco, except
sulfur which was determined by the
Microchemical Analysis Laboratory,
Department of Chemistry, University
of California

The semi-quantitative spectrographic analysis is given in Table 3,
with the reported values being ± 5% of the amount present.
Table 3 also gives the sulfur content, as determined by chemical
analysis.   X-ray powder diffraction analysis, which was carried
out using Cu-K_α radiation, exhibited all the peaks reported for
rutile in the ASTM Powder Diffraction file data.   None of the
peaks reported for anatase, the other polymorphic modification of
TiO_2, were observed.   The B.E.T. surface area of the TiO_2 powder,
as determined by nitrogen adsorption at liquid-nitrogen temperature with the Micromeritics pore volume/surface area analyzer,
was found to be 26 m^2/g.   The surface area determined by the
single-point B.E.T. method using the Stroehlein Area Meter at
liquid-nitrogen temperature was 25.8 m^2/g.

    To study the electrokinetic behavior at one electrolyte
concentration, eight separate mobility measurements were made at
pH values ranging from 4 to 10.   The pH of the suspension was
controlled with sodium hydroxide and nitric acid, and all inorganic
chemicals were reagent grade.   These experiments were carried out
in the absence of ionic strength control so that cations used for
ionic strength control would not be competing with the alkaline-
earth ions being studied.   To insure a good dispersion of the
solids, the suspensions (solid/liquid ratio = 0.01/100) were
given an ultrasonic treatment for two minutes and were subsequently
conditioned for four hours on a wrist-action shaker before conducting the electrophoresis measurements with a Riddick Zetameter.

The adsorption experiments were carried out in polyethylene bottles containing 0.5 g of TiO_2 in 50 mL of solution (1%). The solids were added to a known concentration of metal nitrate salt solutions (Mg, Ca, Sr, Ba) and pH was controlled by NaOH addition. The suspensions were then agitated in a shaker for the predetermined equilibration time after which the solids were separated by filtration and the final pH was measured. The final concentrations in solution were determined by atomic absorption spectrophotometry.

EXPERIMENTAL RESULTS

The first series of experiments was done with sodium nitrate, which has been established to be an indifferent electrolyte for rutile in aqeuous media.[7,8,9,11] Figure 1 shows that the electrophoretic mobility-vs.-pH curves at three different ionic strengths of $NaNO_3$ cross at pH 6.5, which is taken to be the PZC of this sample of rutile.

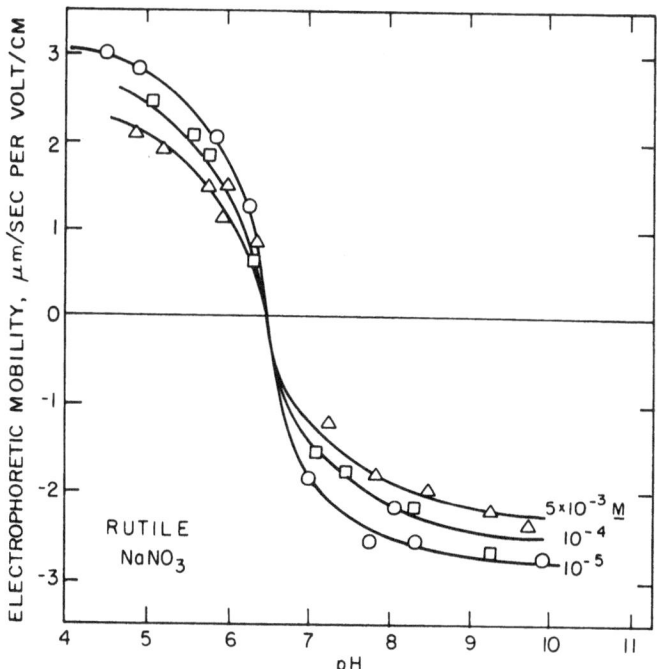

FIGURE 1   Electrophoretic mobility-vs.-pH curves for rutile in the presence of different concentrations of $NaNO_3$. The curves intersect at the PZC.

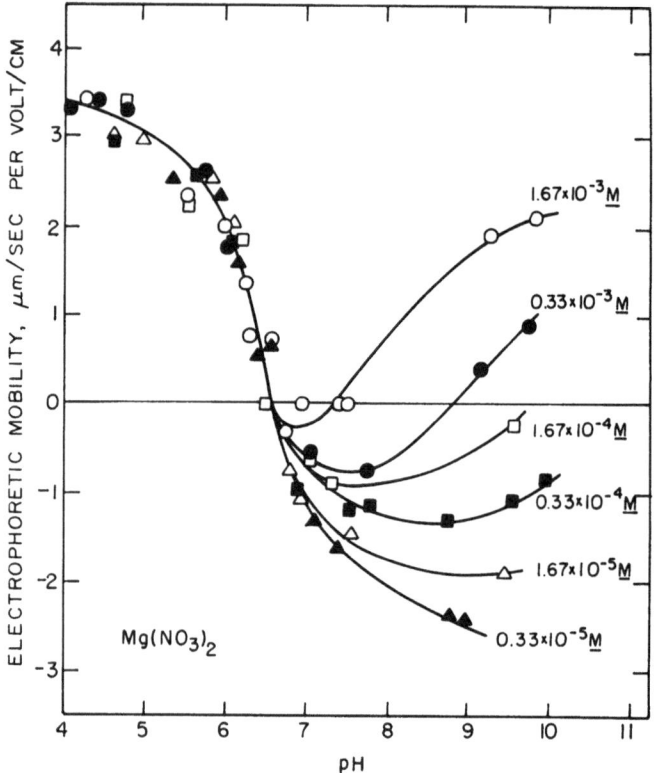

FIGURE 2   Electrophoretic mobility-vs.-pH curves for rutile in
the presence of different concentrations of $Mg(NO_3)_2$.

The electrophoretic mobility studies on rutile in the presence
of alkaline-earth metal ions were carried out using the nitrate
salts of these metals.   Since the nitrate ion functions as an
indifferent counter ion on rutile, any changes observed in electrokinetic behavior can be attributed to the metal ions themselves.
The concentrations of the nitrates for which the electrophoretic
studies were performed were: 1.67 x 10^{-3}, 0.33 x 10^{-3}, 1.67 x 10^{-4},
0.33 x 10^{-4}, 1.67 x 10^{-5} and 0.33 x 10^{-5} M.   Ionic strength was
not controlled by a fixed concentration of an indifferent electrolyte, such as sodium nitrate, because we did not want the results
to reflect ion exchange behavior between Na^+ and the added cation.

Figure 2 shows the mobility-vs.-pH result for rutile in the
presence of $Mg(NO_3)_2$ solutions of different concentration.   For
pH values less than the PZC, the points recorded are scattered
roughly around a single line.   This is, in part, due to retaining
the results in terms of mobilities rather than as calculated zeta
potentials.   For pH values greater than the PZC, the shape of the

mobility-vs.-pH curve depends markedly on the $Mg(NO_3)_2$ concentration.   The results obtained with 0.33×10^{-5} M magnesium nitrate
solution are those typical for an indifferent electrolyte but for
progressively higher concentrations, the mobilities become less
negative as the pH is increased above a certain value.   In fact,
for $Mg(NO_3)_2$ concentrations of 0.33×10^{-3} M and 1.67×10^{-3} M,
there are two pH values at which the mobility changes sign: one
at pH = 6.5 (PZC of rutile) and the other at pH 7.6 or 8.8, for
the respective magnesium nitrate concentration.

The mobility of rutile as a function of pH in the presence
of $Ca(NO_3)_2$ solutions is presented in Figure 3.   The curve for a
$Ca(NO_3)_2$ concentration of 0.33×10^{-5} M is quite similar to the
curve obtained in the presence of $Mg(NO_3)_2$ of the same concentration.   One striking difference between the mobility-vs.-pH results
in the presence of Mg^{2+} and Ca^{2+} occurs at a concentration of
1.67×10^{-3} M.   At this concentration, unlike the $TiO_2/Mg(NO_3)_2$
system, the mobility values in the $TiO_2/Ca(NO_3)_2$ system are
positive in the pH range 4· to 10.   The TiO_2 particles exhibit a
reversal of electrophoretic mobility not only at a pH near the
PZC but also at higher pH's in $Ca(NO_3)_2$ solutions of 0.33×10^{-4} M

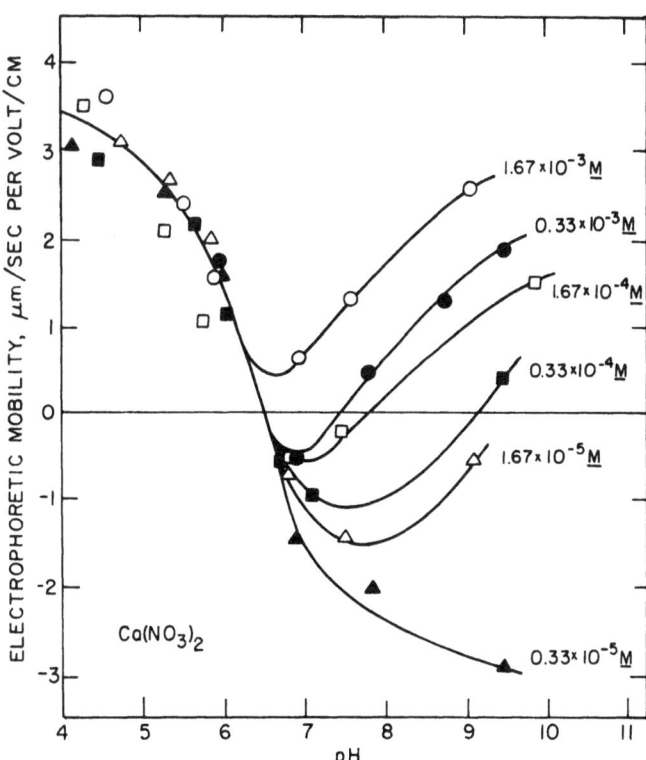

FIGURE 3   Electrophoretic mobility-vs.-pH curves for rutile in
the presence of different concentrations of $Ca(NO_3)_2$.

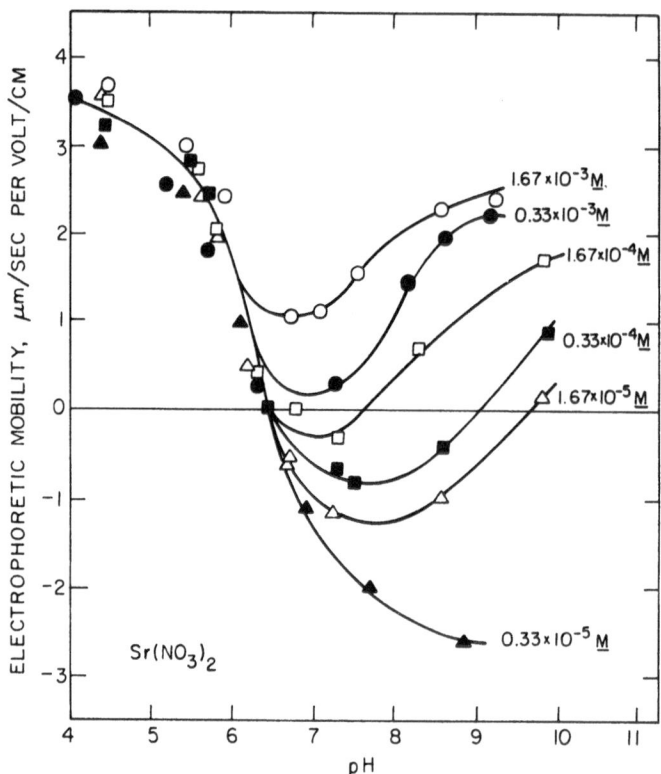

FIGURE 4   Electrophoretic mobility-vs.-pH curves for rutile in
the presence of different concentrations of $Sr(NO_3)_2$.

concentration, or greater.   The decrease in the negative values
of mobility at pH values above the PZC of TiO_2 in the presence of
Ca^{2+}, resulting ultimately in the reversal of charge, is much
more pronounced in the case of Ca^{2+} than Mg^{2+}.

Figure 4, which presents the results obtained for strontium
nitrate solutions, indicates that the tendency for alkaline-earth
ions to reverse the zeta potential of rutile is increased for
Sr^{2+} over Ca^{2+}.

The reversal of the zeta potential of rutile to positive
values in the presence of barium salts, as depicted in Figure 5,
is dramatic.   At all the concentrations studied (from 1.67 x 10^{-3} M
to 1.67 x 10^{-5} M), the particles exhibited positive electrophoretic
mobilities.   Unlike the effect of Mg^{2+} and Ca^{2+} on TiO_2 at a
concentration of 0.33 x 10^{-5} M, Ba^{2+} does not appear to behave
indifferently at this concentration level.   At dosages of Ba^{2+}
above 1.67 x 10^{-5} M, TiO_2 particles appear to be positively
charged in the presence of Ba^{2+} even at pH values lower than the

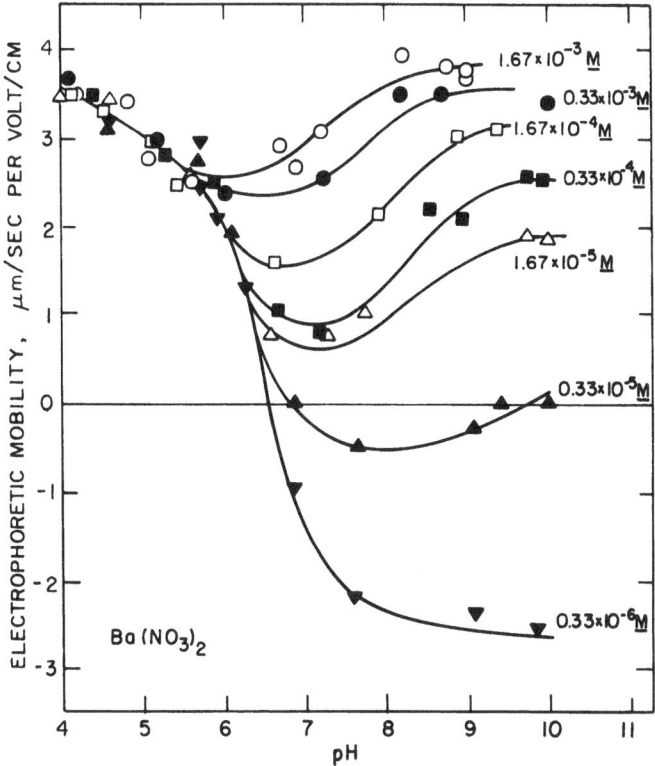

FIGURE 5   Electrophoretic mobility-vs.-pH curves for rutile in
the presence of different concentrations of Ba(NO$_3$)$_2$.

PZC. This indicates that Ba^{2+} is able to adsorb even when the
surface charge of TiO$_2$ is positive.

To corroborate the electrokinetic observations that the
affinity sequence of the alkaline-earth ions follows the order Ba
> Sr > Ca > Mg, a limited series of adsorption experiments was
carried out. For the adsorption tests, 10 grams of rutile per
liter (1% solids by weight) was agitated for six hours (and also
70 hours) in a solution containing 10^{-3} mole of the added nitrate
salt. The pH was maintained at approximately 8.2 with NaOH and
the percentage of cation adsorbed was determined by atomic absorption spectrophotometry. The results are summarized in Table 4.

DISCUSSION

One of the reasons for undertaking this investigation was to
ascertain the effect that the relative size of the alkaline-earth
metals (with respect to each other) would have on their adsorption

Table 4. The Adsorption of Alkaline-Earth Ions on Rutile. Initial concentration of nitrate salt = 10^{-3} M

Cation	6 Hours Equilibration			70 Hours Equilibration		
	Equilib. pH	Final conc., M	% abstracted	Equilib. pH	Final conc., M	% abstracted
Mg	8.1	0.53×10^{-3}	47%	8.2	0.522×10^{-3}	47.8%
Ca	8.2	0.444×10^{-3}	56%	7.6	0.475×10^{-3}	52.5%
Sr	8.3	0.39×10^{-3}	61%			
Ba	8.6	0.23×10^{-3}	77%			

on oxides, and also to study any similarities that may exist because these elements belong to the same group of the Periodic Table.

Thus, it comes to no surprise that some of the explanations that are valid for one set of metal nitrate data are also valid for all the others. On the other hand, some differences that are observed on going from one salt to another can be explained by the fact that these metal cations belong to the same group.

The results will be explained under three headings:

1.   Effect of the alkaline-earth metal ions at pH values less than the PZC.

2.   Effect of the alkaline-earth metal ions at pH values greater than the PZC.

3.   Effect of the ionic radius on adsorption.

1.   Effect of the Alkaline-Earth Metal Ions at pH Values Less Than the PZC

The fact that some adsorption may be taking place on the solid at pH values less than the PZC can be confirmed from three observations:

(i)   The maximum positive electrophoretic mobility value for all the four metal ions is slightly greater than the maximum positive mobility that is obtained with the indifferent electrolyte, $NaNO_3$.

(ii)   For Ca^{2+}, Sr^{2+} and Ba^{2+}, it can be seem from Figures 3, 4 and 5, respectively, that the mobility-vs.-pH plots for higher salt concentrations exhibit positive mobilities at all concentrations tested. This effect is most pronounced for the plot representing 1.67×10^{-3} M barium nitrate, which begins to show a deviation from the curve from other salts at a pH of 5.5. These positive mobility values indicate that some adsorption must be taking place at the PZC, and even at pH's less than the PZC.

(iii)   There were no observable effects of changing the ionic strength with alkaline-earth metal cations on the electrophoretic mobility. From Figure 1, it is apparent that increasing the concentration of an indifferent electrolyte causes a change in the slope of the mobility-vs.-ph curve on both sides of the PZC. However, Purcell and Sun,[5] who used the streaming potential technique, observed a decrease in the zeta potential of rutile at pH's below the PZC as the concentration of calcium was increased.

Perhaps one explanation of the above behavior involves the adsorption of the multivalent metal ions when the surface is positively charged. Another possible way to explain the adsorption of the cation on a positive surface is through the site distribution model.[19] If a model is assumed wherein 50 percent of the oxide surface sites are uncharged at the PZC, cation adsorption can be explained by the electrostatic attraction that may exist between cations in solution and the comparatively large number of negative sites existing at pH's somewhat below the PZC. On the other hand, if it is assumed that there are 80 percent neutral sites at the PZC, the negative sites existing at low pH values are not sufficiently significant to warrant the concentration effects that are observed in the experimental data. That is, by increasing the concentration of the cations, it is observed that the positive deviations are increased. For the kind of deviations that are observed, it is necessary to conclude that there is some other driving force besides electrostatic which makes more cations adsorb on the surface with an increase in the bulk cation concentration.

The specific adsorption of the cations onto a positive surface can explain the fact that the effect of the change in ionic strength is not observed at low pH values. For an indifferent electrolyte, an increase in concentration compresses the double layer, thus reducing the potential at the Stern plate, which reduces the electrophoretic mobility. In the case of specific adsorption, however, the double layer compression is compensated by an increase in potential at the Stern plane because of excess adsorption, and hence there is no change observed in the mobility values.

2.   Effect of Alkaline-Earth Metal Ions at pH Values Greater than the PZC

For pH values higher than the PZC, there are fewer uncertainties as to what might be responsible for the results obtained. According to the site distribution model, the number of negative sites begins to increase almost exponentially with an increase in pH. For every negative site created, the following reactions take place:

for positive sites: $TiOH_2^+ + OH^- \rightleftharpoons TiOH + H_2O$

for neutral sites: $TiOH + OH^- \rightleftharpoons TiO^- + H_2O.$

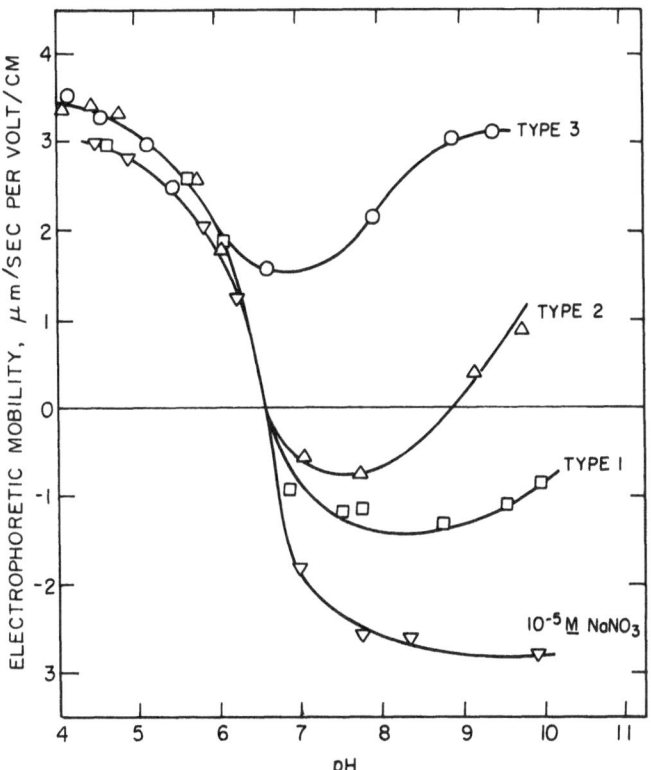

FIGURE 6   Characteristic mobility-vs.-pH plots for alkaline-earth
salts compared to 10^{-5} M NaNO$_3$.

Thus, with increasing hydroxyl concentration (increasing
pH), the negative charge on the solid increases.   The bivalent
cations in the bulk are attracted by this negative surface charge,
and for every negative site that interacts with a metal cation,
there is a net increase of one positive charge:

$$TiO^- + Ba^{2+} \rightleftharpoons TiOBa^+.$$

An explanation of the mobility data above the PZC in the
presence of the various alkaline-earth salts can be made in terms
of the typical plots shown in Figure 6.

For type 1, where a minimum occurs in the negative value of
the mobility, one can interpret the results in the following
manner.   By increasing the pH, the number of negative sites on
the surface is increased, resulting in an increase in cation
adsorption in the Stern plane.   At the left of the minimum, the
increase in the negative charge due to the change in pH is <u>higher</u>

than any increase in positive charge resulting from cation adsorption. At the minimum, the increase in the negative surface
charge is balanced by the adsorbing positively charged cations.
Beyond the minimum, the increase in the negative charge due to
the change in pH is _less_ than the increase in positive charge
resulting from cation adsorption.

For type 2, the number of cations adsorbed is so high that
beyond a certain pH not only do they balance the negative charge,
but they make the net (surface + Stern layer) charge positive.
The pH at which the electrophoretic mobility is zero is called
the point of zeta reversal (PZR). This increased cation adsorption results from higher concentrations of these ions in the
bulk, for any metal under consideration.

For type 3, the cation adsorption is extremely high at all
pH values and so the negative surface charge never exceeds the
positive charge due to adsorbed ions in the Stern layer. It is
worth noting that these graphs also show minima which occur at pH
values around the PZC where the negative sites are the least.

James and Healy[6] found that similar charge reversals occur
for cobalt adsorption on TiO_2 and SiO_2. They explained these
charge reversals in terms of the precipitation of hydrolysis
products of cobalt on the solid surface. Such an argument based
on the hydrolysis of adsorbates prior to adsorption cannot be
invoked in the case of alkaline-earth metal ions as they exhibit
very little tendency to hydrolyze except at pH values above 9, as
is evident from Table 5 which gives the first hydrolysis constants
of the alkaline-earth metal ions. Table 5 presents a summary of
the hydrolysis constants for the alkaline-earth metals and the
concentration of the hydrolyzed species for selected concentrations of the cations in solution. From the table it can be seen
that the most hydrolyzed alkaline-earth cation is Mg^{2+}. It is
also evident from this table that Mg^{2+} hydrolysis takes place
only in moderately basic solutions. For other alkaline-earth
ions hydrolysis is unimportant below a pH of 9. For example, for
Ba^{2+}, the only hydrolyzed species is $[BaOH^+]$ and the concentration
of this species for a barium nitrate concentration of 1.67×10^{-3} M
is 2.45×10^{-6} M at pH 10, whereas for 1.67×10^{-5} M barium
nitrate, the $[BaOH^+]$ concentration at pH 10 is 2.45×10^{-8} M.
Since a charge reversal is observed for both the above concentrations of Ba^{2+}, this clearly rules out metal ion hydrolysis as the
dominant factor in the reversal of the zeta potential on TiO_2.

3.   Effect of the Ionic Radius on Adsorption

From the summary presented in Figure 7, it can be seen very
clearly that for the same bulk ionic concentration of 0.33×10^{-3} M,
there is a definite order for the charge reversal that occurs on

Table 5. Hydrolysis Data for the Alkaline-Earth Metal Cations[50]

Cation	Log10 Equilibrium Constants	Total Conc. of Cation (mol/liter)	Conc. of Unhydrolyzed Cation	Hydrolyzed Species	Conc. of Hydrolyzed Species at pH 10 (mol/liter)
Mg^{2+}	$K_{so} = -10.87$	1.67×10^{-3}*		$Mg(OH)_2$	
	$K_1 = 2.58$	1.67×10^{-3}	1.61×10^{-3}	$[Mg(OH)^+]$	6.1×10^{-5}
		1.67×10^{-5}	3.18×10^{-5}	$[Mg(OH)^+]$	1.2×10^{-6}
Ca^{2+}	$K_{so} = -5.24$	1.67×10^{-3}**		$Ca(OH)_2$	
	$K_1 = 1.32$	1.67×10^{-3}	1.67×10^{-3}	$[Ca(OH)^+]$	3.5×10^{-6}
		1.67×10^{-5}	1.67×10^{-5}	$[Ca(OH)^+]$	3.5×10^{-8}
Sr^{2+}	$K_1 = 0.877$	1.67×10^{-3}	1.67×10^{-3}	$[Sr(OH)^+]$	1.3×10^{-6}
		1.67×10^{-4}	1.67×10^{-4}	$[Sr(OH)^+]$	1.3×10^{-7}
Ba^{2+}	$K_1 = 0.69$	1.67×10^{-3}	1.67×10^{-3}	$[Ba(OH)^+]$	8.2×10^{-7}
		1.67×10^{-4}	1.67×10^{-4}	$[Ba(OH)^+]$	8.2×10^{-8}

NOTE: K_{so} is the $\log_{10}$ of the conventional solubility product.

K_1 is the $\log_{10}$ of the equilibrium constant for the addition of OH^- to the metal ion.

* at this concentration, $Mg(OH)_2$ will precipitate at pH 9.95 or greater.

** at this concentration, $Ca(OH)_2$ will precipitate at pH 12.8 or greater.

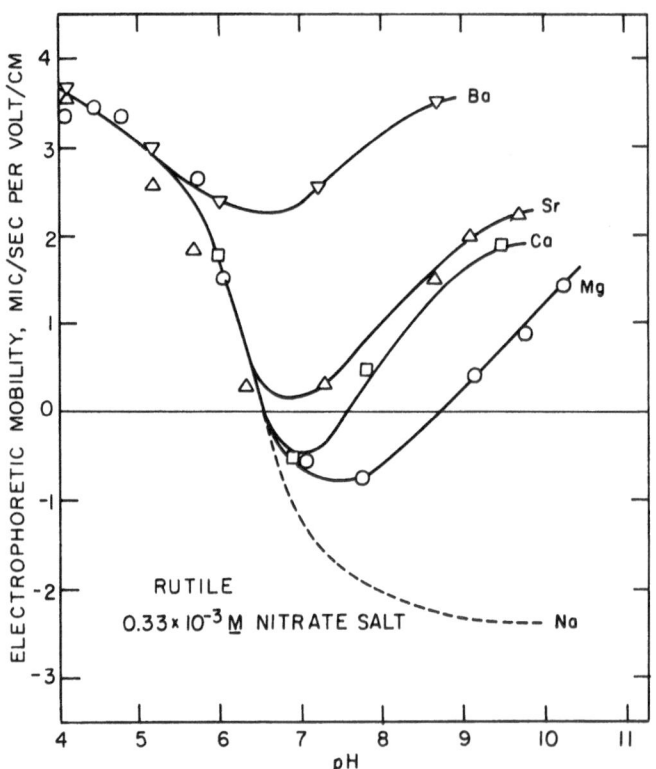

FIGURE 7   Summary of the mobility-vs.-pH data for the different
alkaline-earth cations at 0.33 x 10^{-3} M concentration.

TiO$_2$ particles in the presence of the various alkaline-earth
metals.   If the magnitude of the electrophoretic mobility is
considered to give a measure of the relative magnitude of the
adsorption on the surface, then the maximum change in the Stern
layer charge is brought about by Ba^{2+} ions, and the least by Mg^{+2}
ions.   The order for a given metal ion concentration can be
written as:

    Ba >> Sr > Ca > Mg.

An explanation for the affinity sequence: Ba >> Sr > Ca > Mg,
can be offered in terms of the hydration properties of these
ions.

    In the simplest model of ionic hydration due to Born,[20] the
ions are represented as charged hard spheres.   In water with a
dielectric constant, ε, the hydration energy of an ion with
charge z and radius r', is given by Equation (1) where N is the
Avogadro's number.

$$\Delta G_2 = -\left(\frac{NZ_1^2 e_o^2}{2r_i'}\right)\left[1 - \frac{1}{\varepsilon}\right].$$ (1)

Voet[21] has shown that the hydration energies of alkaline-earth metal ions are consistent with the Born model when r_i' is given by:

$$r_i' = r_1^G + \Delta$$ (2)

where r_i^G is the Goldschmidt crystal radius of the ion and $\Delta = 0.7$ Å for alkaline-earth metal ions. Goldschmidt and Pauling radii are approximately the same for the ions under consideration. Using Equations (1) and (2), the results of the hydration energy are given in Table 6.

From the above equations, we see that the hydration energy is inversely proportional to the radius, and consequently, the smallest ion Mg^{2+} has the highest (negative) hydration energy, and Ba^{2+}, the largest, has the lowest negative hydration energy.

In the case of specific adsorption of a hydrated ion, at least a part of the hydrated shell closest to the surface has to be dehydrated for the ion to be able to go into the Stern plane. To remove these water molecules, work must be done and this work is a function of the hydration energy. The more negative the hydration energy of an ion, the more work is necessary to remove the water. Hence Ba^{2+} ions, which have the least negative hydration free energy, should find it easier to adsorb at the Stern plane compared to Mg^{2+} ions.

The free energy of adsorption, ΔG_{ads}^o, of the cations can be evaluated from the adsorption experiments using the Stern–Grahame equation:

$$\Gamma_\delta = 2 r c \exp\left(-\Delta G_{ads}^o/RT\right)$$ (3)

where Γ_δ is the adsorption density in mol/cm^2, r the radius of the adsorbed ion, c the bulk concentration of the ion in solution in mol/cm^3, R the gas constant and T the temperature in K. The standard free energy of adsorption can be split into its electrical components, $zF\psi_\delta$, and into a specific adsorption component, ΔG_{spec}^o. For these calculations, ψ_δ is taken to be the zeta potential for the given conditions, as evaluated from the electrophoretic mobility studies. Table 7 summarizes these results. Recalculation of the specific adsorption free energy in terms of

Table 6.   Free Energies of Hydration for the
            Alkaline-Earth Metals as Calculated
            from Equation (2)[51]

Ion	r_i' (Å)	$-\Delta G_i$ (kcal/mol)
Mg^{2+}	0.65	455.5
Ca^{2+}	0.99	380.8
Sr^{2+}	1.13	345.9
Ba^{2+}	1.35	315.1

kcal/mol of adsorbed ion shows the following sequence: Mg^{2+},
-6.08 kcal/mol; Ca^{2+}, -6.4 kcal/mol; Sr^{2+}, -7.2 kcal/mol;
Ba^{2+}, -8.6 kcal/mol. This follows the same order as the free
energy for dehydrating these same cations (Table 6). This adds
further corroboration to the postulate that the alkaline-earth
cations dehydrate upon adsorption, and that the relative ease of
dehydration determines the affinity sequence for adsorption in
this system.

There appears to be some indication that the affinity series
is dependent in part on the electrostatic field strength of the
oxide. Oxides with the high field strength, or high heats of
immersion, tend to exhibit an affinity series in the order: Mg >
Ca > Sr > Ba, which is opposite to the order found for most
materials.

ACKNOWLEDGEMENTS

This work was supported by a grant from the National Science
Foundation. The authors wish to thank Mr. Pradip for carrying
out the adsorption experiments and Professor G.D. Parfitt for
extensive discussions about this investigation.

Table 7. Evaluation of the Adsorption Free Energies from Adsorption and Electrokinetic Data

Cation	Ionic radius r, cm	Equilib. bulk conc., c mol/cm^3	Equilib. ads. dens. Γ_δ mol/cm^2	$\Gamma_\delta/2\ r\ c$	ΔG°_{ads}	ψ_δ, mV	$zF\psi_\delta$ ($z = 2$)	ΔG°_{spec}
Mg^{2+}	0.65×10^{-8}	0.53×10^{-6}	1.81×10^{-10}	2.627×10^4	-18.18 RT	0	0	-10.2 RT
Ca^{2+}	0.99×10^{-8}	0.45×10^{-6}	2.14×10^{-10}	2.4×10^4	-10.1 RT	8.1	0.624 RT	-10.7 RT
Sr^{2+}	1.13×10^{-8}	0.39×10^{-6}	2.35×10^{-10}	2.66×10^4	-10.2 RT	24.3	1.874 RT	-12.1 RT
Ba^{2+}	1.35×10^{-8}	0.23×10^{-6}	2.95×10^{-10}	4.75×10^4	-10.8 RT	48.6	3.75 RT	-14.5 RT

REFERENCES

1. D.W. Fuerstenau, Ed., "Froth Flotation - 50th Anniversary
   Volume", AIME, New York (1962).
2. D.W. Fuerstenau and T.W. Healy, Principles of Mineral Flotation, in "Adsorptive Bubble Separation Techniques", R. Lemlich,
   Ed., Academic Press, New York (1972). pp. 92-131.
3. M.C. Fuerstenau, Ed., "Flotation - A.M. Gaudin Memorial
   Volume I", AIME, New York (1976).
4. G.A. Parks, Chem. Rev. 65:177 (1965).
5. G. Purcell and S.C. Sun, Trans. AIME 6 (1963).
6. R.O. James and T.W. Healy, J. Colliod Interface Sci. 40:42,
   53,65 (1972).
7. Y.G. Berube and P.L. De Bruyn, J. Colloid Interface Sci.
   27:305 (1968).
8. Y.G. Berube and P.L. De Bruyn, J. Colloid Interface Sci.
   28:92 (1968).
9. D.N. Furlong and G.D. Parfitt, Powder Tech. 65:548 (1978).
10. T.W. Healy and D.W. Fuerstenau, J. Colloid Interface Sci.
    20:376 (1965).
11. G.D. Parfitt, Prog. Surf. Membr. Sci. 11:181 (1976).
12. A.M. Gaudin and D.W. Fuerstenau, Trans. AIME 202:66 (1955).
13. G.R. Wiese and T.W. Healy, J. Colloid Interface Sci. 51:434
    (1975).
14. A. Breeuwsma and J. Lyklema, Disc. Faraday Soc. 52:324 (1971).
15. D.G. Kenniburgh, Ph.D. Thesis, University of Wisconsin
    (1973).
16. M. Malati and S.F. Estefan, J. Colloid Interface Sci. 22:306
    (1966).
17. Th.F. Tadros and J. Lyklema, J. Electroanal. Chem. 22:9
    (1969).
18. C.P. Huang and W. Stumm, J. Colloid Interface Sci. 43:409
    (1973).
19. R.W. Lai, Ph.D. Thesis, College of Eng., University of
    California, Berkeley (1970).
20. M. Born, Z. Physik 1:45 (1920).
21. A. Voet, Trans. Faraday Soc. 32:1301 (1936).
22. H. Freundlich and Von Elissatoff, Z. Physik Chem. 79:385
    (1912).
23. H.R. Kruyt and M.A.M. Klompe, Kolloid Beih. 54:484 (1942).
24. H. Freundlich, Z. Physik Chem. 44:129 (1903); 73:385 (1910).
25. R. Kunin, "Ion Exchange Resins, 2nd Ed.", John Wiley and
    Sons, New York (1958).
26. A. Breeuwsma, Ph.D. Thesis, Agricultural University, Wage-
    ningen, The Netherlands (1973).
27. J.P. Gabano, P. Etienne and J.F. Laurent, Electrochim. Acta
    10:947 (1965).
28. M. Malati and Gray, J. Chem. Tech. BioTech., in press.
29. J.W. Murray, J. Colloid Interface Sci. 46:357 (1974).

30. M. Malati, R.J. Mazza, A.J. Sherren and D.R. Tompkins, Powder Tech. 9:107 (1974).
31. A. Von Buzagh, Kolloid Z. 47:370 (1929); 51:105,230 (1930).
32. C.B. Amphlett et al., J. Inorg. Nucl. Chem. 26:297 (1964).
33. L.L. Ames, American Mineralogist 45:689 (1960).
34. C.E. Marshall, "Colloid Chem. of the Silicate Minerals", Academic Press, New York (1949). p. 195.
35. H. Freundlich, Kapillarchemie II Leipzig, 387 (1932).
36. J. Lyklema, Trans. Faraday Soc. 59:418 (1963).
37. F. Dumont and A. Watillon, Disc. Far. Soc. 52:352 (1971).
38. T. Grevini and R. Somigliana, Energia Nucl. (Milan) 6:339 (1959).
39. W. Stumm, C.P. Huang and S.R. Jenkins, Croatica Chemica Acta 42:223 (1970).
40. G.E. Boyd, J. Schubert and A.W. Adamson, J. Am. Chem. Soc. 69:2828 (1947).
41. G. Alberti, Atti. Accad. Naz. Lincei., Rend Classe Sci., Fis. Mat. Nat. 31:427 (1961).
42. P. Schachtschabel , Kolloid Beih. 51:199 (1940).
43. C.N. Merriam and H.C. Thomas, J. Chem. Phys. 24:993 (1956).
44. H. Jenny, J. Phys. Chem. 36:2219 (1932).
45. R.M. Barrer, Proc. Chem. Soc. 99 (1958).
46. R.M. Barrer, Chem. Ind. (London) 1258 (1968).
47. M. Malati and A.E. Smith, Powder Tech. 22:279 (1979).
48. F. Dumont and A. Watillon, Paper presented at Euromech 104, Conference in Lourain, Belgium (1978).
49. H.J. Modi and D.W. Fuerstenau, J. Phys. Chem. 61:640 (1957).
50. "Stability Constants of Metal Ion Complexes, Special Pub. 17", The Chemical Soc., London (1964).
51. J. Burgess, "Metal Ions in Solution", John Wiley and Sons, New York (1978).

THE ADSORPTION OF OLEATE FROM AQUEOUS SOLUTION

ONTO HEMATITE

Seng N. Yap, R.K. Mishra, S. Raghavan
and D.W. Fuerstenau

Department of Materials Science
and Mineral Engineering
University of California
Berkeley, California 94720

ABSTRACT

The interaction of sodium oleate with hematite has been
investigated through adsorption, electrophoretic mobility, and
wettability studies. The pH and oleate concentration ranges were
selected to preclude formation of bulk liquid oleic acid or acid-
salt complexes. The effect of pH and temperature on adsorption
and the electrophoretic mobility reveal the existence of chemisorbed oleate at the hematite-water interface. Adsorption mechanisms are postulated and conditions for good flotation response
are predicted. Precautions that should be taken in oleate adsorption experiments using radiotracers are discussed.

INTRODUCTION

The use of oleic acid or sodium oleate in the flotation of
oxide minerals has been practiced for a long time. Although
considerable research has been directed towards the study of
oleate adsorption on iron oxides, there is no absolute agreement
among various investigators about the mechanism of the oleate
adsorption, mainly due to differences in observed experimental
results in this complex system.

The aim of this present work has been to pursue a better
understanding of the nature of adsorption of oleate at the
hematite/aqueous solution interface through adsorption and
electrokinetic measurements. Additional information about the

119

nature of the hematite surface and the mode of oleate adsorption
was obtained from investigations of its extractability by an
organic liquid and from contact angle measurements.

The effect of solid/liquid ratios on adsorption density
which had been observed in some earlier experiments on oleate
adsorption by hydroxyapatite[36] prompted an interest in investigat-
ing the existence of similar effects in the present system. In
most of the experiments, nonradioactive oleic acid converted to
sodium oleate was used as the adsorbate because of its greater
reliability and ease of handling. Some adsorption measurements
were carried out using carbon-14-tagged oleate to delineate the
effects that radioactively tagging the unsaturated oleate chain
might have on the observed results.

BACKGROUND

Previous Oleate/Hematite Adsorption Results

Using infrared spectroscopy, Peck and others[1] discovered
that synthetic hematite, which exhibited a point-of-zero charge
(PZC) at pH 9.0 reacted best with oleic acid at pH 9.4 and with
sodium oleate at pH 7.4. For specular hematite (PZC = pH 7.7),
the maximum adsorption of both oleic acid and sodium oleate
occurred at pH 7.8, and with red hematite (PZC = pH 8.6) the
maximum oleic acid adsorption occurred at pH 8.8 and with oleate
at pH 7.4. Flotation was found to be markedly depressed above
and below these pH values. It was suggested that the reaction
occurs by the following molecular mechanism:

$$M{-}OH + HOL \longrightarrow M{-}OH \ldots\ldots HOL \longrightarrow M{-}OL + H_2O$$

where HOL is oleic acid, M-OH is a mineral surface site with
chemisorbed hydroxyl and M-OL is the mineral surface site with
chemisorbed oleate. Maximum adsorption and flotation at a pH
value near the PZC of the hematite was thought to be attributed
to the maximum number of neutral hydroxyl sites.

Kulkarni and Somasundaran[2] observed that adsorption of
[14]C-tagged oleic acid on hematite increased continuously as the
pH was decreased. No adsorption maximum was found near neutral
pH range though flotation recovery (using nonradioactive oleic
acid) was maximum around neutral pH. A good correlation between
the pH range for best flotation and the highest rate of air/water
interfacial tension decay was observed. They postulated that the
acid soap formed near the neutral pH range may play an important
role in the flotation process.

Using infrared spectroscopy, Paterson and Salman[5] found that
chemisorption of sodium oleate on ferric oxide occurred at pH
values of 8.5 and 10.5. Their adsorption isotherms with ^{14}C-tagged
oleic acid exhibited a single plateau at an adsorption density of
about twice that of a vertically closed-packed monolayer. They
suggested that the chemisorbed species is a dioleate.

Again using infrared spectroscopy, Akhtar[6] found it to be
impossible to determine whether the carboxylate groups of the
oleic acid are directly bonded to surface iron atoms or through
surface -OH groups.

Pope and Sutton[7] found good flotation near neutral pH,
though adsorption was higher at lower pH values. At pH 2, no
flotation was observed even though the adsorption was very high.
At pH 10, flotation ceased at a measured adsorption density of
about 6×10^{-10} mol/cm^2, while at pH 8 flotation still occurred
at higher adsorption densities. Isotherms measured at pH 8 and
pH 10, exhibited two plateaus. Up to the first plateau (Γ =
2.6×10^{-10} mol/cm^2), they suggested that adsorption takes place
by interaction of the carboxylate ion and the olefinic double
bond with the positively charged surface. At higher oleate
concentrations the olefinic group appears to detach from the
surface. When the charge in the adsorbed layer exceeded the
surface charge, further adsorption takes place by hydrocarbon
chain interaction with the carboxylate group pointing to the
aqueous phase (giving the second plateau of 7.6×10^{-10} mol/cm^2).
This increased the hydrophilicity of the surface and flotation
eventually ceased.

Gutierrez[8] observed that maximum adsorption of ^{14}C-tagged
oleate on hematite and ilmenite occurred in a narrow pH range
(near neutral pH) when the concentration of the oleate was low
and the conditioning time was short. This narrow maximum in the
adsorption-vs-pH curves disappeared at higher concentrations and
longer adsorption time. Maximum adsorption at pH 7.2 - 7.6 was
attributed to the fastest rate of adsorption under those conditions. Gutierrez in a later publication[9] showed a good correlation between oleate adsorbed and hematite flotation at low oleate
concentrations. However, at oleate concentrations greater than
5×10^{-5} mol/L at acid pH's, although the adsorption of oleic
acid was very high, flotation was very poor. This was thought to
be attributed to the disordered character of the layer of cooper-
atively adsorbed oleic acid.

Generally iron oxides have been found to have a great affinity
towards oleic acid or sodium oleate. However, LaPointe,[11] found
that the adsorption of ^{14}C-tagged oleic acid on hematite did not
exceed 12 percent of monolayer coverage. Due to the reversibility
of the adsorption, he concluded that the adsorption was purely a
physical phenomenon.

The diverse results obtained by various investigators have demonstrated the need of further study for achieving a better understanding of the nature of oleate adsorption at the iron oxide/aqueous interface. These diverse observations are probably attributed, among others, to the complicated behavior of aqueous oleic acid (oleate) systems, differences in experimental materials and differences in experimental conditions and techniques. Since knowledge of the properties of aqueous solutions of oleic acid/ oleate and the use of appropriate experimental techniques are essential for the study of oleate adsorption, it should be useful to briefly review the properties of aqueous oleic acid systems and the techniques generally used in the adsorption studies.

The Properties of Aqeuous Oleic Acid/Sodium Oleate Solutions

The properties of oleic acid and its aqueous solutions have been reported in a number of publications.[12,13] A diagram exhibiting the relative predominance of various sodium oleate species in water [Figure 1, after Moon[25]] is useful for determining conditions for the study of oleate adsorption. The domain region of each oleate species as well as the change of the boundaries can be easily observed in this diagram. The quantity $[Na^+]$ is the total sodium ion activity resulting from sodium oleate and any added indifferent electrolytes. The $NaHOL_{2(\ell)}$ phase begins to appear only when $[Na^+]$ is greater than $10^{-2.78}$ in the absence of added salt. Upon increasing the amount of sodium in the system, the domain for $NaHOL_{2(\ell)}$ expands. As the total $[Na^+]$ concentration is increased, also the boundary between OL^- and OL_2^{2-} (line 3) moves from pH 8.55 to higher pH values. With increasing total oleate concentrations, the boundaries between $NaHOL_{2(\ell)}$ and OL^- (line 5), OL_2^{2-} (line 7), micelles (line 11) move towards higher pH values while the boundary between $NaHOL_{2(\ell)}$ and $HOL_{(\ell)}$ (line 6) moves towards a lower pH region. Also, the relative concentrations of each oleate species for a total oleate concentration of 10^{-10} molar are shown in Figure 1, as an example.

In adsorption measurements where the disappearance of oleate species from solution after the addition of solid is assumed to be the amount adsorbed, conditions which permit bulk precipitation of the oleate species or the formation of liquid HOL and $NaHOL_2$ should be avoided. Figure 1 can be used as a guide for selecting appropriate experimental conditions.

Various Techniques for Measurement of Oleate Adsorption

As briefly summarized, such techniques as infrared spectroscopy, differential thermal analysis, UV spectrophotometry, and radiotracer analysis have been used to study or measure oleate

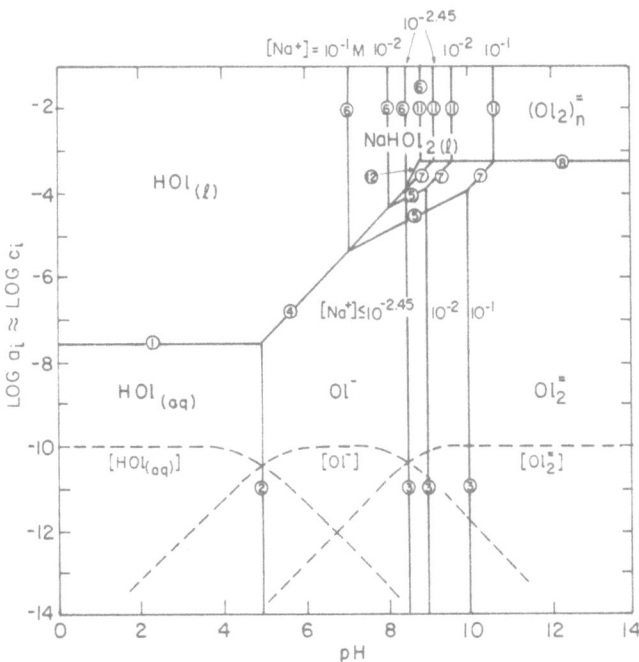

FIGURE 1   Domains of oleate species in aqueous solutions as a
            function of pH at 25°C. The effect of Na^+ on the
            domains is also illustrated. The dashed lines show the
            distribution of oleic acid, oleate ions, and oleate
            dimers as a function of pH at a constant total oleate
            addition (after Moon[25]).

adsorption. Radiotracer techniques have been most widely used
for simplicity and high sensitivity. Despite these advantages,
however, the radiotracer technique suffers disadvantages which
should not be overlooked.

First of all, it has been assumed tacitly that the radioactive oleic acid behaves exactly like the untagged oleic acid,
but this assumption may not be correct. For instance, Newman[28]
observed an interfacial isotope effect in the measurements of the
stability of Ca-H-St monolayers (stearic acid monolayer on subsolu-
tions containing calcium). He observed a sudden, unexplained
decrease in the stability of the ^{14}C-tagged Ca-H-St monolayers
above pH 7.8, which was not observed in the measurements using
untagged stearic acid.

Secondly, in aqueous solutions, oleic acid decomposes slowly
in the presence of small amounts of oxygen[29] but in the presence
of ^{14}C, the decomposition takes place more rapidly.[30,33]

Presumably, the beta particles may interact with water molecules
to produce hydroxyl radicals which in turn initiate the decomposition of the olefinic double bond of oleic acid. Unfortunately,
no work has yet been conducted to investigate the effect of ^{14}C
on the rate of decomposition/oxidation of oleic acid under the
conditions used in adsorption or flotation experiments, though
the effects of higher energy beta and gamma radiations on the
collecting properties of oleic acid and sodium oleate have been
observed.[31,32]

Finally, in contrast to the spectrophotometric method, the
radiotracer technique is unable to differentiate between ^{14}C-oleic
acid and its soluble decomposition products. Therefore, the
radioactive solutions should be periodically analyzed by a spectrophotometric or chromatographic method parallel with counting even
when no visible change is observed to ascertain the extent of any
decomposition. Decomposition phenomena can drastically affect
the reliability of experimental results or adsorption behavior.

On the other hand, nonradioactive oleic acid offers a number
of advantages over the ^{14}C-tagged oleic acid, such as in handling.
An aqueous solution of nonradioactive oleic acid (HOL) can be
handled in the presence of trace amounts of oxygen without significant decomposition within 12 hours. In contrast, ^{14}C-tagged
oleic acid (due to its faster decomposition in the presence of
oxygen), should be used under oxygen-free conditions.

As for storage and solution preparation, oleic acid can be
transformed and crystallized as sodium oleate and stored in a
dark bottle at room temperature for practically an indefinite
time even without excluding the air from the container. On the
other hand, ^{14}C-HOL should not be stored as the solid sodium
salt, but as oleic acid diluted with an inert non-aqeuous solvent
(preferably containing double bonds such as benzene) under nitrogen or vacuum*. After several months of storage, the ^{14}C-HOL
even if stored in benzene, must be purified prior to use. Storage
of the sodium salt provides an advantage in solution preparation
owing to the ease of sodium oleate dissolution in slightly basic
aqueous solutions. On the other hand, the preparation of ^{14}C-HOL
aqueous solutions is time consuming in that the organic solvent
must be evaporated first, and then the dissolution of oleic acid
is very slow even at pH 10 or slightly greater.

---

*A comprehensive review of self-decomposition (and methods of
  storage to reduce the decomposition of ^{14}C-labeled organic
  compounds) has been given by Rochlin[33] and of long-chain compounds
  by Bayly and Evans.[34]

Compared to the other analytical techniques, infrared spectroscopy has the advantage of being able to differentiate between
the physical and chemical adsorption. However, IR methods are
open to criticism in that KBr pellet preparation techniques may
in themselves lead to surface compound formation. Although,
Gregory's UV spectrophotometric method[37] only measures the amount
of oleate that disappears from the solution rather than the
amount adsorbed, it is relatively simple and has higher sensitivity
and better accuracy compared to infrared and differential thermal
analysis. The experimental error is less than 1% for oleate
concentrations of 1.5×10^{-5} mol/L or greater, and less than 8%
for concentrations near 3×10^{-6} mol/L. If conditions which
permit precipitation of liquid HOL and $NaHOL_2$ can be avoided, the
use of Gregory's spectrophotometric method to determine adsorption
densities should be quite satisfactory.

EXPERIMENTAL MATERIALS AND METHODS

The synthetic hematite used was reagent-grade material
obtained from Baker Chemical Company. Its specific surface area
was 7.9 m^2/g as determined by the B.E.T. method, using nitrogen
as the adsorbate.

Sodium oleate was prepared by titrating a hot alcoholic
solution of 99+% pure oleic acid (obtained from NU-CHEK-PREP.
Inc. Minnesota) with a freshly prepared alcoholic solution of
analytical grade NaOH using phenolphthalein as indicator, then
recrystallized in hot alcohol. The sodium oleate containing
radiotracer was prepared in the same way using ^{14}C-tagged oleic
acid in benzene solution (obtained from ICN; 99 mg in 3 mL of
benzene, specific activity = 10 m Ci/mmol) diluted 60-fold with
pure oleic acid. The scintillator liquid was prepared by mixing
42 mL Tolu Scint I with Spectragrade dioxane to make 1000 mL
solution (both reagents were obtained from ICN). Twelve mL of
the scintillator liquid was used for each 1.00 mL radioactive
aqueous solution.

The water used throughout the experiments was triply distilled and purged with nitrogen filtered through a CO_2 absorbent.

The adsorption experiments were carried out in 52 mL-glass
stoppered centrifuge tubes calibrated to ± 0.5% of its volume.
Known amounts of hematite were introduced into the tubes, followed
by freshly prepared sodium oleate solutions of known concentration
and pH. The supporting electrolyte was 2×10^{-3} mol/L $NaNO_3$.
The suspensions were dispersed in an ultrasonic bath for 30 s and
were then equilibrated in a dark chamber by rotating the tubes
parallel to the longitudinal axis. After centrifugation, the
residual concentrations were determined by the spectrophotometric

method described by Gregory.[37] For oleate concentrations below
8×10^{-6} mol/L more aqueous phase was taken for the analysis (2.5
times) while the organic phase was maintained the same. Blank
tests analyzed in triplicate were carried out for each set of
experiments. The difference between the blank and the residual
concentration was taken as the amount adsorbed. The ^{14}C-oleate
concentrations were determined by liquid scintillation counting.

For the high-temperature studies, the oleate solutions were
preheated to the desired temperature before mixing with the
hematite. The ultrasonic bath was heated to the same temperature,
± 2°C. The equilibration was conducted in a constant temperature
bath obtained from the Bayley Instrument Co.

The pH and the oleate concentration ranges were chosen such
that there was no turbidity due to oleic acid nor acid-salt
complex, based on the phase diagram of the relative predominance
of various oleate species (Figure 1).

The adsorption reversibility test was conducted by replacing
a known amount of clear solution of an equilibrated suspension by
2×10^{-3} mol/L NaNO$_3$ solution of the same pH and then agitated
for 4 hours. The solutions before and after dilution were analyzed.
The electrophoretic mobility of hematite at various oleate concentrations and pH was measured with a Zeta-Meter, using a solids
concentration of 0.04%.

Before conducting the adsorption experiments, the rate of
the decomposition of sodium oleate in solution was studied at 23°
and 58°C. Within the pH range studied, no significant decomposition of the oleate species was found within 12 hours at 23°C, and
9 hours at 58°C; no loss of activity nor turbidity was observed
in the radioactive oleate blank solutions within 24 hours at
23°C. To ensure no decomposition of oleate had occurred, the
dissolution of the sodium oleate and the analysis were conducted
the same day.

The hydrophobic or hydrophilic character of the hematite
surface after adsorption of oleate was observed by shaking the
suspensions with cyclohexane. Instantaneous complete extraction
of the particles into the oil phase or accumulation at the oil/
water interface is considered to be hydrophobic, while the non-
extractability by the oil phase is considered to be hydrophilic.
If the material was not extractable by oil at first but extractable after prolonged shaking, the hematite was considered to be
hydrophilic due to reverse orientation of the oleate at high
coverage; prolonged shaking would remove this physically adsorbed
oleate leaving the oleate adsorbed with the tail towards the
aqueous phase and at that condition, hydrophobic.

EXPERIMENTAL RESULTS

    The results presented here were obtained with nonradioactive
sodium oleate, unless otherwise mentioned.

Rate of Adsorption

    In order to determine the appropriate conditioning time for
the adsorption studies, a series of experiments were conducted to
measure the amount of oleate adsorbed on hematite as a function
of time at two pH values and three different temperatures with an
initial oleate concentration of about 2.6×10^{-5} mol/L. The
results showed that the adsorption is very rapid with equilibrium
being reached in less than 3 minutes at pH 9.2 at 23°C and 60°C
and also at pH 9.8 at 60°C. At pH 9.8 at 23°C and pH 9.2 at 3°C,
even though the adsorption is less, equilibrium is essentially
attained in less than 20 minutes, with more than 85% being adsorbed
within 3 minutes. On this basis, 3 to 4 hours conditioning time
was used for experiments conducted at 23°C or lower and two hours
for higher temperatures to ensure that adsorption equilibrium was
attained even at low oleate concentrations and higher pH values.

    No significant decomposition of oleate is detected within 12
hours at 23°C and 8 hours at 60°C.

Adsorption Isotherms

    When hematite is dispersed in an aqueous solution, the pH of
the solution decreases significantly; the magnitude depends upon
the initial pH, the amount of solids, and the concentration of
sodium oleate. Therefore, to determine the adsorption isotherms,
it is more practical to measure first the amount of oleate adsorbed
as a function of pH at various oleate concentrations and use
these data to construct the adsorption isotherms.

    Measurements of the uptake of oleate as a function of pH at
23°C at three different levels of solids additions with various
initial oleate concentrations were carried out. Figure 2 presents
isotherms for the adsorption of oleate by hematite at various pH
values. Clearly, there is no solid/liquid ratio effect in this
system. All of the isotherms tend to a plateau at an adsorption
density of about 10^{-9} mol/cm^2. If this is considered to be a
closed-packed monolayer, the oleate would have a cross-sectional
area of about 17 Å^2/molecule. However, the cross-sectional area
of a vertically oriented carboxylic acid is about 20 Å^2, as
determined from film pressure-area isotherms.[38] Therefore, at
the plateau the adsorption appears to be more than a closed-
packed monolayer. The vertical lines on the isotherms in Figure 2

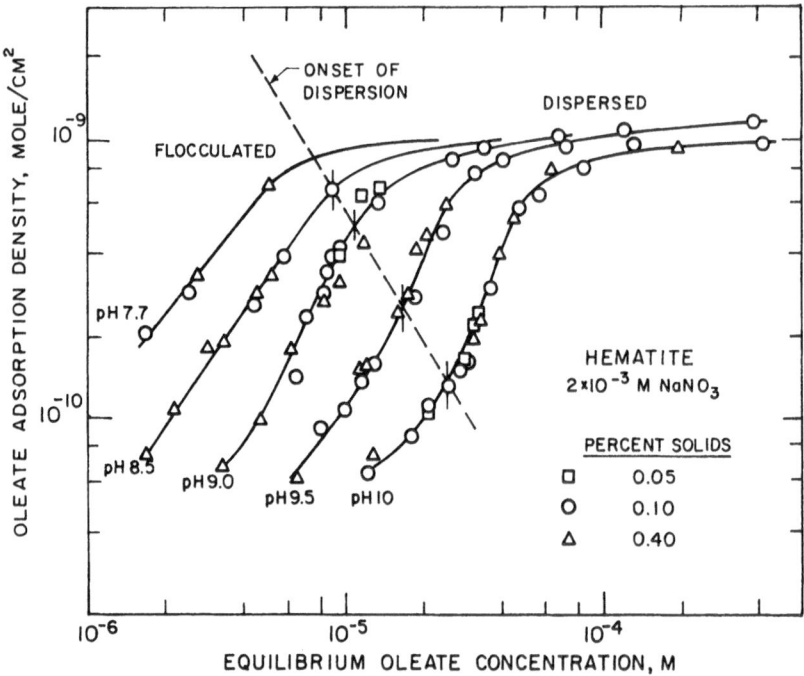

FIGURE 2   Adsorption isotherms of oleate on hematite in 2 x 10^{-3}
            mol/L $NaNO_3$ at different pH values at 23°C. The regions
            of flocculation and dispersion are also shown.

mark the oleate concentration below which all the solids are
flocculated. At that particular concentration one begins to
observe dispersion of some of the particles and the amount of
solids dispersed increases with increasing oleate concentration.
At the plateau the solids are very well dispersed. Also, the
particles remain dispersed in the aqueous phase after shaking for
about one minute with cyclohexane, indicating that the solids are
hydrophilic. After shaking for a longer time, however, the
material becomes hydrophobic and is completely extracted into the
organic phase. It is interesting to note that with increasing pH
values the solids start to disperse at lower surface coverage.
When the solids are flocculated, under most conditions they can
be extracted immediately and quantitatively into the organic
phase (or be collected at the droplet surface), indicating that
the solids are strongly hydrophobic.

        The dependence of adsorption density on pH at various oleate
equilibrium concentrations is illustrated in Figure 3. At very
low oleate concentrations (i.e. 4 x 10^{-6} mol/L), the adsorption
decreases markedly as the pH is increased (above pH 7). At
higher oleate concentrations, the oleate uptake remains at the

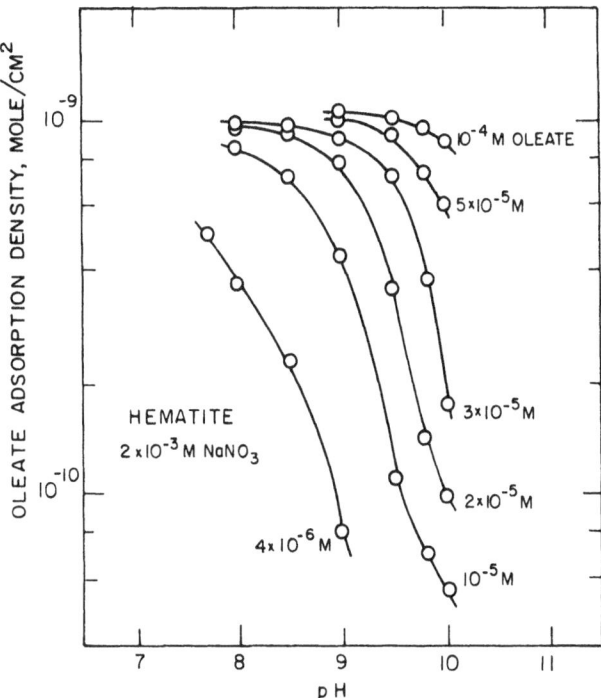

FIGURE 3   The adsorption density of oleate on hematite as a
           function of pH at various equilibrium concentrations of
           sodium oleate in 2 x 10^{-3} mol/L $NaNO_3$ and 23°C.

plateau until the pH is increased considerably above the PZC of
the hematite, that is above pH 8.2.   A significant amount of
oleate adsorption is observed even at pH values of 1 or 2 pH
units above the PZC, indicating the occurrence of chemisorption.
No adsorption maximum at the PZC of the hematite was observed, as
found by Peck.[1]   This is in agreement with the findings of
Kulkarni,[2] Pope[7] and Paterson.[5]

Adsorption of ^{14}C-Tagged Oleate

The adsorption of ^{14}C-tagged oleate on hematite was found to
be considerably lower than that of untagged oleate conducted
under the same conditions.   This is apparently due to the significant decomposition of the oleate as indicated by the spectrophotometric analysis of the radioactive blank solutions, although
neither turbidity nor loss of activity was observed in the blank
solutions.   Our results, however, were very similar to those
reported by Kulkarni[2] if the adsorption isotherm is plotted for
pH 9.2 (1 pH unit above the PZC) when compared to his results
conducted at pH 8 (the PZC of his hematite was at pH 7).

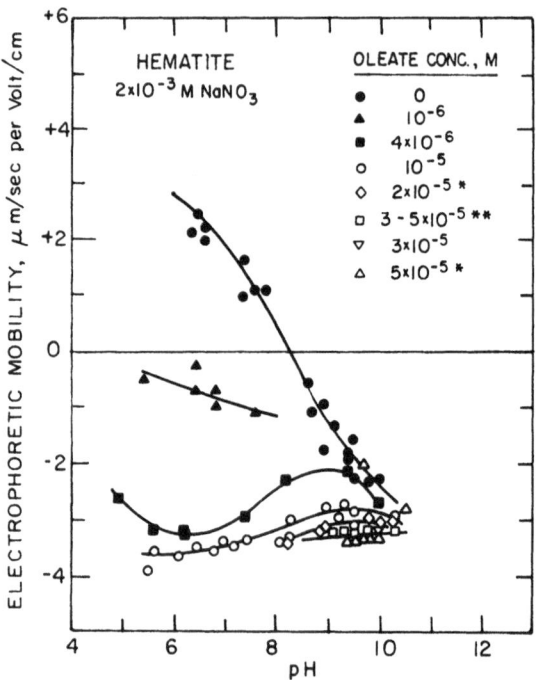

FIGURE 4   The electrophoretic mobility of hematite as a function
           of pH at various sodium oleate additions in 2 x 10⁻³
           mol/L NaNO₃, 23°C and 0.004% solids by weight.
           *These data points are the mobility of hematite taken
            from the adsorption experiments.
           **These data are for the mobility of hematite taken
            from adsorption experiments with an initial oleate
            concentration of 9.5 x 10⁻⁵ mol/L but 0.1% solids.

         In one test at pH 9.8 to study the reversibility of adsorp-
tion the solution was diluted and after desorption the expected
equilibrium amount of oleate was achieved, indicating reversibility
of adsorption under these conditions.

Electrophoretic Mobility

         The electrophoretic mobility of hematite as a function of pH
in the presence of various oleate concentrations is presented in
Figure 4. A very low solids content was used (0.004%) so that
the oleate equilibrium concentrations would be close to the
initial concentrations. Some mobility measurements were also
conducted using the suspensions from the adsorption experiments.
As can be seen in Figure 4, the PZC of the hematite occurs at
about pH 8.2, in agreement with the previous results.[39,40]

Clearly, sodium oleate has a marked effect on the electrokinetic
potential of hematite since a large negative mobility was observed
at pH 8.2 even at a concentration as low as 4×10^{-6} mol/L.
Actually the mobility was negative under all oleate concentrations
and pH ranges studied. Chemisorption seems to occur even at this
low concentration.

Effect of Temperature on Adsorption Density

The adsorption density of oleate on hematite at 23, 43, and
60°C as a function of pH is illustrated in Figure 5. These
results indicate that the adsorption density is independent of
temperature in the range 23 - 60°C, which is in agreement with
the results of Kulkarni.[2] Other experiments showed that adsorption
decreases when the temperature was decreased from 23° to 10°C.

Since temperature below room temperature appeared to affect
the adsorption behavior, more detailed study was carried out at
pH 9 at 23, 10, and 3°C. The results, which are plotted in
Figure 6, show that adsorption definitely decreases with reduction
in temperature, indicating either an endothermic reaction or the
occurrence of an activated adsorption reaction at the surface.

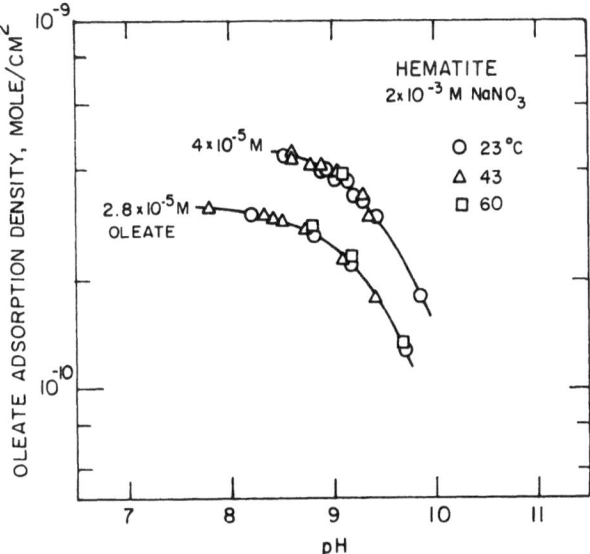

FIGURE 5   Adsorption density of oleate on hematite as a function
           of pH at three different temperatures, 0.1% solids,
           with initial concentratons of sodium oleate of 2.8×10^{-5}
           and 4×10^{-5} mol/L in 2×10^{-3} mol/L NaNO$_3$.

Contact Angle Measurements

Using the free-bubble contact angle method with a disk
sintered by hot pressing the hematite powder, contact angles were
evaluated at 23°C. Figure 7 presents the contact angle of hematite
as a function of pH for 10^{-6}, 10^{-5}, and 10^{-4} mol/L oleate. These
results indicate the same sort of relationship between contact
angle and pH and equilibrium oleate concentration as was observed
with adsorption behavior. At 10^{-4} mol/L oleate, the maximum
contact angle is somewhat less, indicating possible reverse
orientation at higher coverages.

DISCUSSION OF RESULTS

In this paper, oleate uptake will be discussed in terms of
primary and secondary adsorption. Secondary adsorption entails
the uptake of a second layer of oleate through hydrocarbon chain
interaction between the attached oleate monolayer and the additional oleate ions. Primary adsorption is considered to be oleate
uptake in the first layer which can occur through any or all of
the following types of interactions: electrostatic, chemical
bond formation, hydrogen bonding and chain-chain interactions
(hydrophobic bonding). The effect of solution conditions on
these adsorption mechanisms will be discussed, and wettability of
the hematite will be interpreted in terms of adsorption behavior.

Primary Adsorption

From study of the infrared spectra of oleic acid and sodium
oleate on hematite, Peck[1] and Paterson[5] concluded that the adsorption involves chemical bond formation between surface iron atoms
and the carboxylate groups of the oleate species. Akhtar's work,[6]
however, indicated that from infrared spectra it was not possible
to determine whether the carboxylate groups are directly bonded
to a surface iron atom or through the surface OH groups. Due to
the reversibility of the adsorption, other investigators believed
that the adsorption process is a physical one.[9,10] The present
investigation has shown that increasing the temperature from 3°
to 23°C increases the adsorption rate and the amount adsorbed.
These observations together with the fact that significant adsorption occurs even at pH values 1 - 2 pH units above the point-of-
zero-charge of the hematite clearly indicates that the adsorption
of oleate on hematite is not purely physical but quite likely is
a combination of chemisorption and physisorption.

The adsorption and the electrophoretic mobility data indicate
that the change in zeta potential in this system is not always
related to the amount adsorbed. In most systems involving the

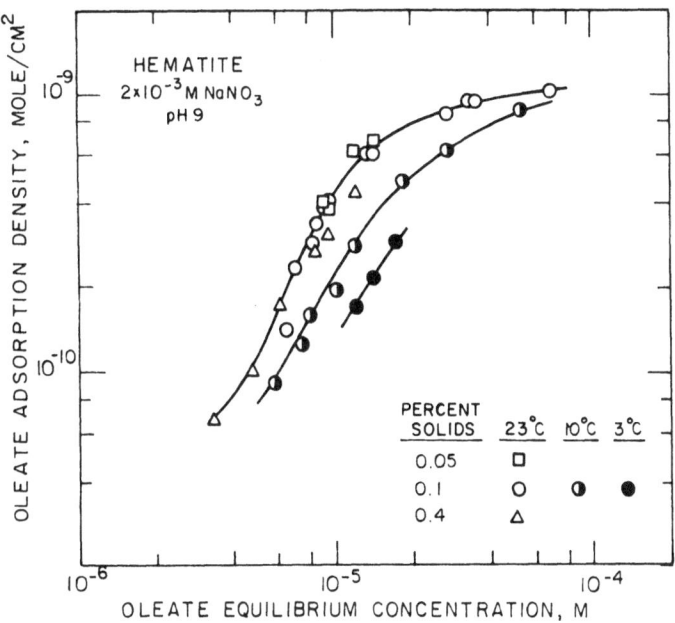

FIGURE 6   Adsorption isotherms of sodium oleate on hematite at pH
9.0 in 2 x 10^{-3} mol/L NaNO$_3$ at 23, 10, and 3°C.

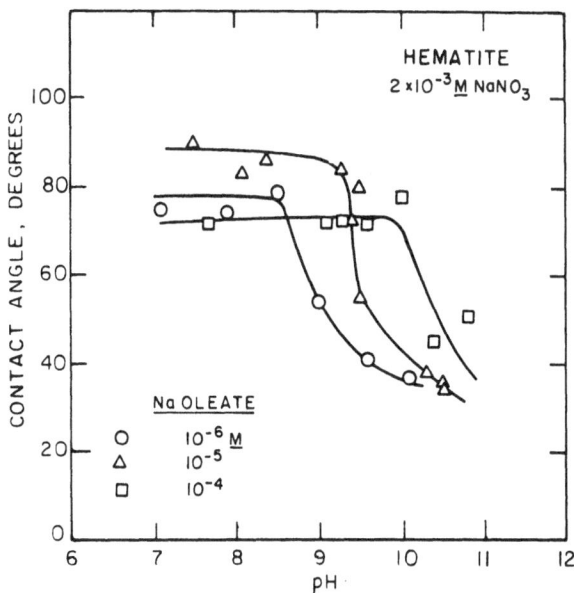

FIGURE 7   The contact angle of hematite as a function of pH at
three different sodium oleate concentrations.

adsorption of ions, there are changes in the zeta potential, which can be interpreted in terms of the Stern-Grahame equation, as shown by Somasundaran and Fuerstenau.[45] At low pH values the adsorption results in large zeta-potential changes, while at high pH values only a small change is observed (Figure 4). Recalling that the PZC of this hematite occurs at pH 8.4 and that 2×10^{-3} mol/L $NaNO_3$ is present, large changes in the zeta potential at pH's below pH 8 would indicate strong specific adsorption of oleate ions which must also be displacing chloride ions from the double layer. It is possible that at high pH values, some of the adsorption is accompanied by the liberation of negative surface species, such as surface hydroxyl groups during the chemisorption process.

The interaction between oleate ions and surface sites may take place by several possible types of mechanisms.

a) Electrostatic interaction between the oleate ions and the positive surface sites

b) Interaction between the olefinic double bond and the positive sites

c) Displacement of the water molecules (positive site) by oleate ions

If there is a hydroxyl group bonded to the same iron as is illustrated above, the formation of a chelate increases the strength of the adsorption.

d) The oleate ion hydrogen bonded to the surface neutral sites

e) Displacement of a surface hydroxyl group by an oleate ion

   This chemisorption reaction (e) may be the predominant
adsorption mechanism at higher pH values, since adsorption does
not change the zeta-potential significantly. This reaction
should occur less readily than reaction (c), because the formation
of H_2O is energetically more favorable than OH^-. This is also in
accordance with the observation that adsorption of oleate proceeds
less readily as the pH increases. At lower pH values, a lower
oleate concentration is needed to produce the same amount of
adsorption because under these conditions there are more positive
surface sites which would favor electrostatic interaction and the
chemical reaction. Of course, total adsorption would depend on
how oleic acid molecules interact at the surface. As the pH
increases, the amount of positive sites decreases while the
negative sites and the $[OH^-]$ increase; as a result, both the
electrostatic and the chemical reaction become less favorable.
At very high pH values, the high negative surface charge and the
high $[OH^-]$ cause the adsorption and the reaction to become very
unfavorable.

   According to the foregoing mechanisms, if they are all
operative, no adsorption maximum would be expected to occur at a
narrow pH range around the PZC of the hematite, as observed by
Peck but in agreement with the present observations.

   The reaction product between the oleate ion and the surface
iron in the pH range studied is likely to be a monooleate rather
than a diolate as suggested by Paterson, because the major reaction
product of oleate and Fe(III) at pH 7.5 is $Fe(OL)(OH)_2 \cdot mHOL \cdot nH_2O$.[41]
Only at lower pH values $Fe(OL)_2OH \cdot mHOL \cdot nH_2O$ is obtained.

   For one set of conditions, the Stern-Grahame equation was
used to assess the magnitude of the free energy of adsorption
ΔG°_{ads} :

$$\Gamma_\delta \;=\; 2\, r\, c \, \exp\left(\frac{-\Delta G^\circ_{ads}}{RT}\right)$$

where Γ_δ is the adsorption density in mol/cm^2, r is the effective
radius of the adsorbed ion (5.2×10^{-8} cm for the carboxylate

ion), R is the gas constant, and T is the absolute temperature
(296 K).   Since the zeta potential is constant at −42 mV (EM −3.3
μm/(s/V/cm)), all of the adsorbed oleate must be in the Stern
plane.   In these calculations

$$\Delta G^\circ_{ads} \;=\; zF\psi_\delta + \Delta G^\circ_{spec}$$

where z is the valence of oleate (−1), F is the Faraday constant
and ψ_δ is approximated by the zeta potential (−42 mV).   Hence the
electrical contribution to the adsorption free energy is 1.66 RT.
These calculations were carried out for a single adsorption
density, namely 7 x 10^{-10} mol/cm^2 and are summarized in Table 1.
ΔG°_{spec} involves the contribution to adsorption from chemical
interaction, hydrocarbon chain association, hydrogen bonding,
etc. as given in reactions (a) to (e).   As can be seen in the
results summarized in Table 1, the standard free energy of specific interaction (which includes all of the above mentioned contri-
butions) is not constant but decreases somewhat as the pH is
increased.   This indicates that some definite but not large
change must be occurring in the nature of the adsorption process
as the pH is increased.

Secondary Adsorption
────────────────────

        The postulated primary adsorption mechanism does not account
for the dispersing action of the oleate species.   According to
Paterson's findings, dispersion started at an adsorption density
slightly greater than monolayer coverage.   It was suggested that
the dispersion is due to the formation of a reversed orientated
second layer.   In this study, however, it was found that at
higher pH values, the dispersion may start at a coverage considerably below a vertically closed-packed monolayer, with adsorption density at the onset of dispersion decreasing with increasing pH.   Because the zeta potential does not change significantly
under these conditions (Figure 4), dispersion is not merely
caused by the formation of a second layer with the polar head
towards the aqueous phase, but rather, may be due to some combined
effect of the zeta potential and the adsorption density of a
reverse oriented oleate species at the first layer.   Examination
of Figure 1 leads one to consider whether the increase in the
dispersing action of the oleate species at higher pH values is
due to the increasing $[OL_2^{2-}]$.   As can be seen in the figure, at
pH 8.55 the $[OL^-] = [OL_2^{2-}]$, if $[Na^+] \leq 10^{-2.45}$.   The $[OL_2^{2-}]/[OL^-]$
ratio continues to increase as the pH increases.   It is believed
that in OL_2^{2-} the hydrocarbon chains of the OL^- are attached to
one another with the polar heads being oriented away from each
other.[42]

Table 1. Evaluation of the Standard Free Energy of
Adsorption of Oleate on Hematite for an
Adsorption Density of 7 x 10^{-10} mol/cm^2
at 23°C

pH	Equilibrium Concentration mol/cm^3	$\left[\dfrac{\Gamma_\delta}{2rc}\right]$	ΔG°_{ads}, RT	ΔG°_{ads}, kcal/mol	ΔG°_{spec}, RT	ΔG°_{spec}, kcal/mol
7.7	5.2 x 10^{-9}	2.59 x 10^{6}	-14.77	-8.69	-16.43	-9.66
8.0	7 x 10^{-9}	1.92 x 10^{6}	-14.47	-8.51	-16.13	-9.48
8.5	1 x 10^{-8}	1.35 x 10^{6}	-14.11	-8.30	-15.77	-9.27
9.0	1.6 x 10^{-8}	8.41 x 10^{5}	-13.64	-8.02	-15.3	-8.99
9.5	3 x 10^{-8}	4.49 x 10^{5}	-13.40	-7.65	-15.06	-8.62
9.8	5 x 10^{-8}	2.69 x 10^{5}	-12.50	-7.35	-14.1	-8.32
10.0	6 x 10^{-8}	2.24 x 10^{5}	-12.32	-7.24	-13.98	-8.22

The dimer OL_2^{2-} can be adsorbed with one or both heads attached
to the surface. The first mentioned will make the surface hydrophilic; and since with increasing pH the ratio of the $[OL_2^{2-}]$ to
$[OL^-]$ increases, for the same coverage the hydrophilicity would
increase as the pH increases. This is likely to be the reason
for the ceasing of flotation at monolayer coverage at pH 10, but
not at pH 8 as observed by Pope. But then, a question arises:
If this mode of adsorption is correct, then the adsorption of the
dimer might increase the negative zeta potential in accordance
with the amount of reversed oriented or physically adsorbed
oleate ion. However, the zeta potential was nearly constant at
high pH values. Possibly, adsorption takes place with the formation of either an ion pair between the carboxylate group and Na$^+$
or with the hydrolysis of the carboxylate group oriented towards
the aqueous phase such that an adsorbed dimer has one polar head
oriented towards the surface and the other hydrolized to COOH.
This is analogous to the NaHOL$_2$ which is formed at high oleate
concentration around pH 9 (Figure 1). This might account for the
near constancy of the zeta potential.

At pH values where the OL$^-$ is the predominant species, the
hydrophilicity of the surface at the plateau is likely due to a
second layer of reversed orientated oleate. After the primary
adsorption of the oleate species (or after the surface charge is
balanced and neutral sites reacted), further adsorption of OL$^-$
takes place through hydrocarbon interaction, and would have the

polar heads away from the surface and the surface would become
hydrophilic at about monolayer coverage.

The Role of pH and Adsorption on Wettability

Comparison of the contact angle curves (carried out with a
sintered hematite disk) given in Figure 7 with the adsorption
curves given in Figure 3 shows that the upper pH limit for forming
a hydrophobic surface is related to the decrease in oleate
adsorption caused by the surface charge becoming highly negative.
The lower maximum contact angle observed at higher oleate concentrations, as well as the initial hydrophilicity observed in the
hexane extraction experiments, probably results from the second-
layer of reverse oriented oleate ions.

In agreement with the results of Kulkarni, our present
results show no adsorption maximum at the PZC of the hematite,
although we did not work much below the PZC because of the oleate
solution chemistry. The difference between Peck's observation,
which showed maximum adsorption near the PZC, with others perhaps
results from the difference in the experimental conditions used
in the experiments. Peck worked only under conditions where
$HOL_{(\ell)}$ or $NaHOL_{2(\ell)}$ are the predominant species where oleate
adsorption is likely to proceed through a molecular mechanism, as
described earlier. In addition, Peck measured the adsorption
after rinsing the solid several times to remove the physically
adsorbed oleate while others measured the total amount adsorbed.
The dramatic increase in adsorption at lower pH values observed
by Kulkarni and Pope is likely to be the result of deposition of
the $HOL_{(\ell)}$ at the surface of the solid. In the present investigation, the conditions which permitted the formation of $HOL_{(\ell)}$
and $NaHOL_{2(\ell)}$ were avoided.

As shown in Figure 1, the composition of the various oleate
species varies with pH and total oleate concentration; that is,
at pH 8 at an oleate concentration of about 10^{-5} mol/L, OL^-
is the predominant species. At pH 8.55, for $[Na^+] \leq 10^{-2.45}$,
$[OL^-] = [OL_2^{2-}]$, and at pH's greater than 8.55, OL_2^{2-} becomes the
predominant species. (At the surface, the boundary for $[OL^-]$
$= [OL_2^{2-}]$ may be shifted to higher pH values since the $[Na^+]$ at
the surface should be higher.) When the pH is decreased to about
7.4, $HOL_{(\ell)}$ starts to precipitate. The formation of $HOL_{(\ell)}$ will
affect the kinetics of chemisorption, especially at low pH values
where the surface may be armored by a rigid water structure due
to the high surface charge and thereby kinetically retard chemisorption. At pH values > 8.5, the increase may contribute to the
decrease in flotation recovery. Therefore, the flotation is
likely to be optimum at a pH of about 8, a prediction in agreement
with the results reported by Kulkarni where flotation recovery

was maximum near pH 7 - 8. On the other hand, Kulkarni[2] has
suggested that the high recovery around neutral pH range was
attributed to acid-salt complex which was formed near neutral pH.
At high ionic strength this may occur but does not seem possible
for low oleate concentrations and low ionic strength, where the
amount of the acid-salt complex is insignificant.

Effect of Solid/Liquid Ratio on Adsorption Density

No significant effect of solid/liquid ratio on adsorption
density is observed under all experimental conditions studied.
Examination of the adsorption rate showed that the adsorption is
very fast at higher pH values (i.e., pH 9.8), though the adsorption rate is slower, more than 80% is adsorbed within the first
2-3 minutes. Apparently, flocculation occurs after the adsorption
equilibrium is attained or is nearly attained. This observation
supports the opinion suggested by our work with apatite[36] that
the effect of solid/liquid ratio observed under conditions of low
zeta potential and slow adsorption rate results from adsorption
kinetics being slower than flocculation kinetics.

Effect of Temperature on Adsorption

The effect of temperature on the adsorption of oleate
by hematite is quite complex. Figure 5 shows that temperatures
above room temperature have little or no effect on the adsorption
process. In Figure 5, the adsorption as a function of pH at two
different oleate concentrations is given for 23, 43 and 60°C.
However, as the temperature is reduced below room temperature,
adsorption decreases. In Figure 6, the results are plotted for
adsorption at pH 9 as a function of oleate concentration at 23,
10 and 3°C. This increase in adsorption with increasing temperature (below room temperature) indicates either an endothermic or
an activated adsorption process. At this stage, in the absence
of calorimetric data, this very complex behavior does not warrant
detailed analysis.

SUMMARY AND CONCLUSIONS

Studies on the adsorption of sodium oleate at the hematite/
water interface at various pH's and temperatures indicate the
existence of a process involving chemisorption together with
physical adsorption at higher coverages. Adsorption and wetting
phenomena are interpreted in terms of the double layer at the
hematite surface and the complex chemistry of aqueous oleate
solutions. To produce a hydrophobic hematite surface, acidic

conditions where liquid oleic acid is formed and highly basic
conditions where oleate dimer is the predominant species should
be avoided. At high adsoption densities the surface is hydrophilic due to the adsorption of reverse-oriented oleate. This
condition is a complex function of oleate concentration, pH, and
the PZC of the hematite.

No effect of solid/liquid ratio on adsorption density was
observed in this system, most probably because of the fast adsorption rate in comparison to the flocculation rate. Conditions for
the flocculation and dispersion of hematite in aqueous oleate
solutions appears to be a complex function of adsorption coverage,
pH and oleate solution chemistry.

Study on oleic acid adsorption using radiotracer techniques
should be conducted with extra precautions, apparently due to the
faster decomposition of the oleic acid in the presence of high
radioactivity. Analysis of the oleate aqueous solutions by a
technique specific for oleate species or sensitive to chain-
length should always be conducted in addition to counting.

Temperature has a complex effect on adsorption in that above
room temperature adsorption appears to be independent of temperature whereas below room temperature adsorption decreases with
decreasing temperature.

ACKNOWLEDGEMENT

The authors wish to acknowledge the National Science Foundation
for the support of this research.

REFERENCES

1. A.S. Peck, L.H. Ruby and M.E. Wadsworth, Trans. AIME 235:301
   (1966).
2. R.D. Kulkarni and P. Somasundaran, Colloids and Surfaces
   1:387 (1980).
3. P. Somasundaran and R.D. Kulkarni, Paper presented at 103rd
   AIME meeting, Dallas (1974).
4. R.D. Kulkarni and P. Somasundaran, AIChE Symposium Series
   71:150,124 (1975).
5. J.G. Paterson and T. Salman, Trans. IMM 79:C91 (1970).
6. S. Akhtar, U.S.N.T.I.S., No. 235015/6GA, PB Rep., 33 (1973);
   in C.A. 82:160670q (1975).
7. M.I. Pope and D.I. Sutton, Powder Tech. 7:271 (1973).
8. C. Gutierrez, paper submitted to Trans. AIME.
9. C. Gutierrez and J. Iskra, International J. Mineral Processing
   4:163 (1977).

10. E.J. Parkins and H.L. Shergold, Flotation, V. 1, A.M. Gaudin Memorial Volume, M.C. Fuerstenau, Ed., AIME, New York, 561 (1976).

11. C.M. LaPointe, Can. Dept. Mines Tech. Surv., Mines Branch Res. Rept. R 108, 33 (1963).

12. P.M. Solozhenkin and others, Dokl. Akad. Nauk. Tadzh. SSR 19:7,31 (1976); in C.A. 86:4,19916a (1977).

13. A.M. Abeid, Indian J. Technol. 14:15,249 (1976).

14. O.S. Bogdanov and N.S. Mikhailova, Obogasshch. Rud. 11:4,19 (1966); in C.A. 66:87700n (1967).

15. N.S. Mikhailova, Tr. Vsed. Nauchn.-Issled, i Proektn-Inst. Mekhan. Obrabotki Polezn, Iskop., 134:31 (1964); in C.A. 64:4665f (1966).

16. V.S. Korobkov and others, Fiz, Khim, Technol., Sb. Mater. Nauch.-Tekh. Konf. Rab. Nauki Proizvod., 174 (1968); in C.A. 75:39089j (1971).

17. S.R.B. Cooke, I. Iwasaki and H.S. Choi, Trans. AIME 214:920 (1959).

18. S.R.B. Cooke and others, Trans. AIME 217:76 (1960).

19. C. Gutierrez, International J. Mineral Processing 3:3,246 (1976).

20. J. Sliwiok and T. Kowalska, Revue Roumaine de Chemie 16:3,439 (1971).

21. R. Roos, J. Oil Technol. Assoc. India 3:1,27 (1971).

22. T. Kowalska, J. Sliwiok, Zesz. Probl. Postepow, Nauk. Roln. 136:123 (1973); in C.A. 79:32891p (1973).

23. B. Stainsby and A.E. Alexander, Trans. Faraday Soc. 45:585 (1949).

24. B. Tamamushi, M. Shirai and K. Tamaki, J. Chem. Soc. Japan 31:467 (1958).

25. K.S. Moon, unpublished research progress report, Dept. of Materials Science and Mineral Engineering, University of California, Berkeley, California (1976).

26. J. Powney, Trans. Faraday Soc. 34:363 (1938).

27. M.A. Cook, J. Phys. Colloid. Chem. 55:383 (1951).

28. R.D. Newman, Nature 250:725 (1974).

29. P.O. Fredrikson and R. Marcuse, Fat Oil Chem., Scand. Symp., 4th, 1965, 147 (published 1967).

30. Personal communication with ICN.

31. I.N. Plaskin and others, Tsvet. Metal 41:9,26 (1968); in C.A. 70:13568m (1969).

32. P.M. Solozhenkin, Dokl. Akad. Nauk. Tadsh. SSR 9:8,28 (1966); in C.A. 66:12559q (1967).

33. P. Rochlin, Chem. Rev. 65:685 (1965).

34. R.J. Bayly and E.A. Evans, J. Labeled Compounds 2:1 (1966).

35. M. Muramatsu, in "Surface and Colloid Science", V. 6, Egon Matijevic, Ed., John Wiley and Sons, New York, 101 (1973).

36. S.N. Yap, R.K. Mishra and D.W. Fuerstenau, in preparation.

37. G.R.E.C. Gregory, Analyst 91:251 (1966).
38. A.W. Adamson, "Physical Chemistry of Surfaces", 2nd ed.,
     Interscience Publishing Co., New York, 158 (1960).
39. K.N. Han, T.W. Healy and D.W. Fuerstenau, J. Colloid Interface
     Sci. 44:3,407 (1973).
40. S. Raghavan and D.W. Fuerstenau, J. Colloid Interface Sci.
     50:2,319 (1974).
41. I.A. Vainshenker and E.D. Kriveleva, Obogashch. Rud. 15:4,64
     (1970); in C.A. 74:132608s (1971).
42. P. Mukerjee, K.J. Mysels and C.I. Dulin, J. Phys. Chem.
     62:1390 (1958).
43. I. Iwasaki, S.R.B. Cooke and H.S. Choi, Trans. AIME 217:237
     (1973).
44. J. Iskra, C. Gutierrez and J.A. Kitchener, Trans. IMM. 82:C73
     (1973).
45. P. Somasundaran and D.W. Fuerstenau, J. Phys. Chem. 70:90
     (1966).

ADSORPTION OF POLYACRYLAMIDE AND SULFONATED

POLYACRYLAMIDE ON Na-KAOLINITE

A.F. Hollander, P. Somasundaran
and C.C. Gryte

School of Engineering and Applied Science
Columbia University
New York, New York 10027

ABSTRACT

The role of solution properties such as pH, ionic strength, temperature, and solid-to-liquid ratio in determining the adsorption of polymers on minerals is studied by conducting tests using labeled polymers and homoionic kaolinite. Data on the effects of the above variables on the adsorption of polyacrylamide (PAM) and sulfonated polyacrylamide copolymers (PAMS) are discussed in terms of changes in polymer charge, configuration and hydrolysis and surface properties of kaolinite in an attempt to develop mechanisms governing the adsorption of such polymers on minerals in aqueous solution.

INTRODUCTION

Polymers are used in a number of industrial processes, involving flocculation, water clarification, filtration, mobility control, etc. In many of these processes, their performance is governed, to a large measure, by the nature and extent of adsorption of the polymers on various mineral particulates. Adsorption is, in turn, expected to depend on such system properties as pH, ionic strength, and temperature. The effect of these parameters on the adsorption of polymers has been studied for a few systems[1-21] but there has been no detailed investigation of the role of all the relevant system variables in the adsorption of nonionic and ionic polymers on well characterized mineral adsorbents. Because of the complex macromolecular nature of the polymers, their

143

adsorption is much more complicated than that of inorganic or
surface active ions on solids. Adsorption from aqueous solution,
in all cases, is governed by the interactions between the solid
and the adsorbate which, in turn, are influenced by their interactions with the solvent. In the case of the polymers, an additional major factor is the configuration of the polymer molecule
and the changes in it during, as well as subsequent to, the
process of uptake. Alterations in configurations of adsorbed
polymer molecules can lead to important changes in the fraction
of polymer segments in active contact with the surface of solids.
All such changes can, in turn, lead to further variations in
adsorption density and in the performance of interfacial processes
such as flocculation.

The above changes in the adsorption density also make it
difficult to use conventional approaches for the theoretical
evaluation of adsorption isotherms. Freundlich and Langmuir
isotherms have been found to describe adsorption in certain
polymer/solid systems[14-20] but to fail in general when applied to
a wide range of concentrations and solution conditions.[17,21]
The loop-type adsorption of polymer segments has been taken into
consideration in the theoretical treatment of isotherms by Simha
et al.[22-24] and Silberberg[25-28] but neither of these isotherms
applies for polymers differing in adsorption energy. Even though
there exists a large number of theoretical isotherms, it appears
that, in order to test and identify isotherms which are valid for
mineral substrates, there is a need for data on adsorption of
different polymers on various minerals under a wide range of
solution conditions. Our investigation therefore covers this
wide range of conditions and polymers and minerals and we also
present our results for the adsorption of nonionic polyacrylamide
and anionic polyacrylamide sulfonate copolymers on Na-kaolinite.

EXPERIMENTAL

Polymers

Polyacrylamide (PAM) and acrylamide/2-acrylamido-2-methyl-
1-propane sulfonic acid copolymer (PAMS) used in this study were
prepared by the precipitation polymerization technique of Wada et
al.[29]

$$\left[\begin{array}{l} -CH_2 - CH - \\ \qquad\quad | \\ \qquad\quad C = O \\ \qquad\quad | \\ \qquad\quad NH_2 \end{array} \right]_n$$

(PAM)

$$\left[\begin{array}{l} -CH_2 - CH - \\ \qquad\quad | \\ \qquad\quad C = O \\ \qquad\quad | \\ \qquad\quad NH \\ \qquad\quad | \\ H_3C - C - CH_3 \\ \qquad\quad | \\ \qquad\quad CH_2 \\ \qquad\quad | \\ \qquad\quad SO_3^- \\ \\ \qquad\quad H^+ \end{array} \right]$$

(Sulfonated unit of PAMS)

Details of the procedure have been described elsewhere.[30]
Briefly, it consisted of irradiation of the ^{14}C labeled acrylamide
solutions in acetone/water mixtures using a cobalt-60 irradiation
source and freeze-drying of the polymer products after removing
the unreacted monomer by washing with acetone. Sulfonated copolymer
was prepared similarly using a mixture of acrylamide, tagged
acrylamide and 2-acrylamido-2-methyl-1-propane sulfonic acid.
Unlike the emulsion polymerization technique, the radiation
initiation method permitted preparation of polymers that are not
contaminated by surfactants. The polymers were characterized by
intrinsic viscosity measurements using an Ubbelohde capillary
viscometer. Molecular weights estimated using the Mark-Houwink
method are 1.18×10^6 for PAM and 1.08×10^6 for PAMS. Radius of
gyration of the PAM is calculated to be 570 Å and the end-to-end
distance to be 1400 Å. Elemental analysis indicated the PAMS to
contain 6.8 percent of the sulfonic acid monomers.

The ionic character of the polymers is best tested by determining the intrinsic viscosity of the polymer solution as a
function of ionic strength. An ionic polymer is expected to show
a greater sensitivity to changes in ionic strength. In Figure 1,
which shows the intrinsic viscosities of the two polymers as a
function of the ionic strength, it can be seen that only the PAMS
is sensitive to ionic strength confirming that it is ionic in
character and that the PAM is essentially nonionic. As expected,
at high ionic strengths both polymers show similar intrinsic
viscosities since the repulsion between charged groups and swelling
of the charged polymer segments due to osmotic pressure gradients
will be minimized by the added salt under such conditions.

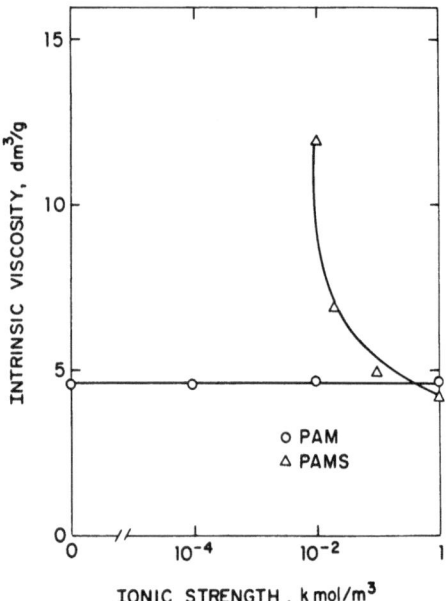

FIGURE 1 Intrinsic viscosities of polyacrylamide and sulfonated
          polyacrylamide as a function of the ionic strength of
          the solution.

      The extent of possible polymer degradation during adsorption
conditions was determined since any such degradation could have
its own effect on adsorption. These tests showed that while the
polymer did not undergo an appreciable change during dry storage
or during storage in the solution form at room temperature, there
was considerable alteration, as indicated by the marked increase
in viscosity, under high temperature conditions particularly when
the solution was agitated (see Figure 2). It is to be noted that
such an increase in viscosity did not occur under high ionic
strength conditions where, possibly due to the retracted configuration, the polymer molecules are not subjected to shear to the
same extent as under low ionic strength conditions. There was
some degradation of the polymer in alkaline solutions but again
the degradation was indicated to be minimal in salt solutions
(see Figure 3). From these results it appears that the effect of
polymer degradation can be controlled for the present tests under
all pH and temperature conditions by adding sufficient salt.

Kaolinite

      Homoionic Na-kaolinite was prepared from a University of
Missouri repository sample of a well crystallized Georgia kaolinite by repeated washing with NaCl solutions using the procedure

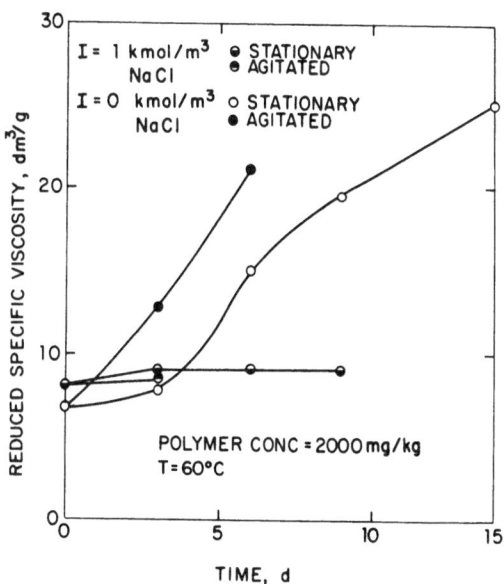

FIGURE 2   Diagram illustrating changes in reduced viscosity of
polyacrylamide with aging of the solutions under higher
temperature conditions (60°C).

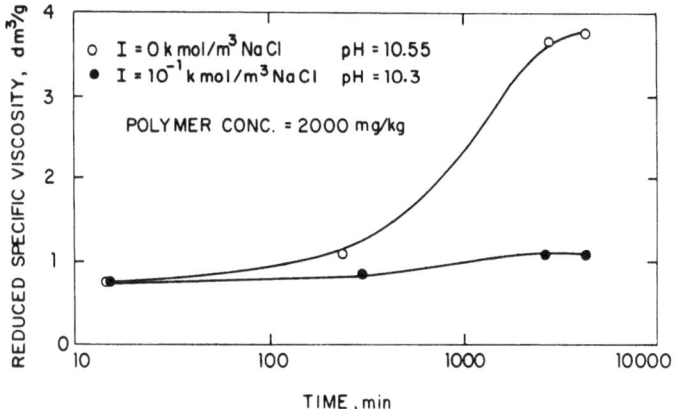

FIGURE 3   Diagram illustrating changes in reduced viscosity of
polyacrylamide in alkaline solutions.

described in Figure 4.   The surface area of the resultant clay
determined by the nitrogen adsorption technique was 9.82 m^2/g.

Experimental Procedure

For adsorption tests, the clay was first equilibrated with
the salt solution for two hours and then polymer solution was
introduced keeping pH and ionic strength constant and agitated
for the desired interval.   All the tests were conducted in a
thermostated incubator.   The ratio of solid to liquid was maintained at 1:10 by weight.   At the end of the agitation, a sample
of the solution was centrifuged and the polymer concentration in
the supernatant was determined by the liquid scintillation counting
technique.   Adsorption density was calculated from the difference
between the initial and final polymer concentrations.   In the

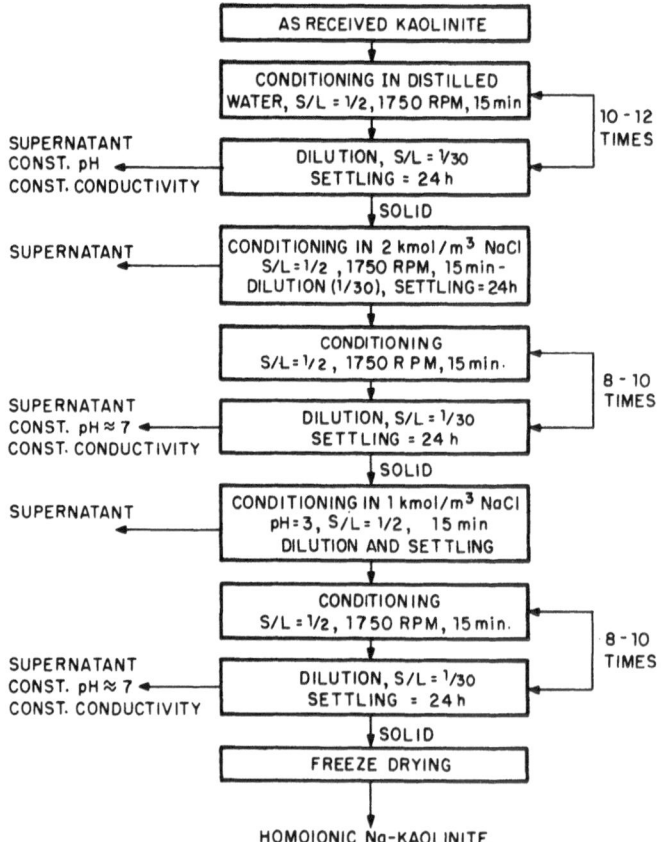

FIGURE 4   Flowsheet of the procedure used for the preparation of
            homoionic Na-kaolinite.

case of the tests at 60°C, the samples were allowed to settle at
60°C for a period of twenty-four hours before centrifugation.
During the 10-minute centrifugation, the temperature decreased by
8°C. Settling at 60°C prior to centrifugation was designed
towards minimizing the effect, if any, of the decrease in temperature during centrifugation.

RESULTS AND DISCUSSION

Adsorption Kinetics

The kinetics of adsorption of polyacrylamide on Na-kaolinite
is illustrated in Figure 5. There is some initial rapid adsorption of the polymer followed by the uptake of additional polymer
at a lower rate. Interestingly, equilibrium adsorption density
is attained at lower polymer concentrations within a shorter
period. It appears that there are at least two processes controlling adsorption in this case, diffusion-controlled adsorption
resulting in initial rapid uptake followed possibly by a slower
rearrangement of polymer molecules already adsorbed on a particle
permitting adsorption of additional polymer on its surface. Such
rearrangement promoted by the crowding polymer molecules is not
expected to be prominent at low polymer concentrations since the
solution is more extensively depleted during the first stage
itself under such conditions. The results in Figure 5 clearly
support the above consideration. In addition, adsorption during
prolonged conditioning of the mineral with the polymer can also

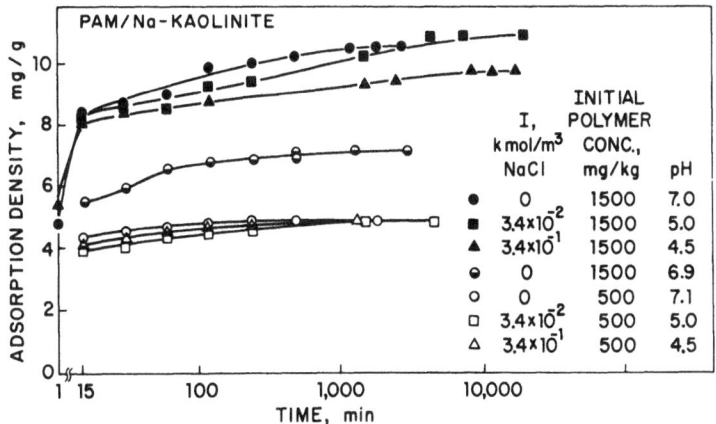

FIGURE 5   Adsorption of polyacrylamide on Na-kaolinite as a
function of time around neutral pH conditions.

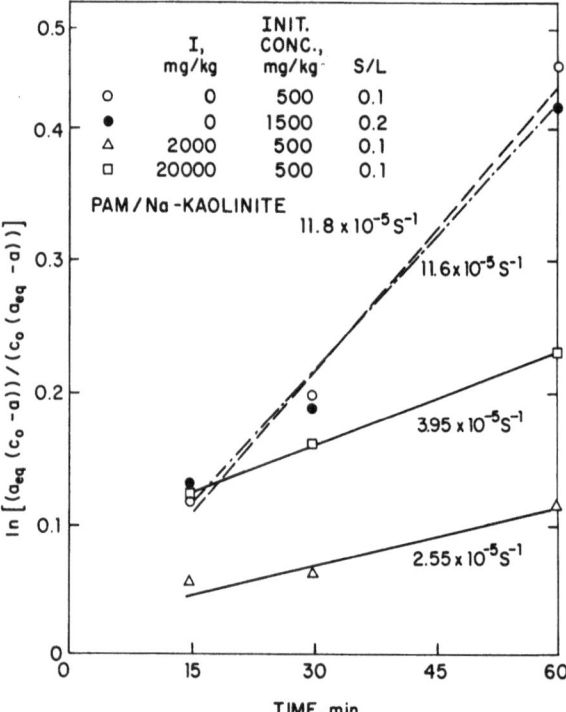

FIGURE 6   Plot of $\ln(a_{eq}(c_o-a)/c_o(a_{eq}-a))$ versus adsorption time.

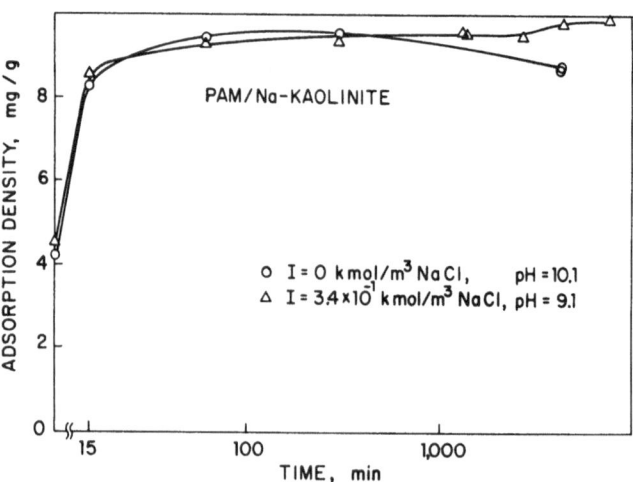

FIGURE 7   Adsorption of polyacrylamide on Na-kaolinite as a function of time under alkaline pH conditions.

be expected to be influenced by the slow dissolution of kaolinite.
The influence of increase in ionic strength and solid/liquid
ratio is also illustrated in Figure 5. Increase in ionic strength
is found to increase the time required to reach equilibrium. The
effect of these parameters on the rate of uptake is estimated by
treating the adsorption data according to Langmuir's rate theory
as developed by Jankovics[7] for cases where desorption rate is
close to zero and the final polymer concentration is only a small
fraction of the initial concentration. Figure 6 shows Jankovics'
plot for the present data where $\ln(a_{eq}(c_o-a))/c_o(a_{eq}-a)$ is plotted
as a function of time and a_{eq}, a, and c_o are respectively equilibrium adsorption density, adsorption at the time t and concentration. The slope of this plot yields the adsorption rate. The
adsorption data treated in the above manner clearly show that the
adsorption rate is reduced markedly by the addition of salt and
the accompanying pH decrease due to both a retracted polymer
configuration and an increased acidity of kaolinite in salt
solutions. The polymer chains can be expected to be less extended
in salt solutions and therefore likely to exhibit a reduced rate
of adsorption. Also clay will undergo reactions with added salt
that can lead to continued adsorption of the polymer. Kaolinite
structure is composed of alternate sheets of alumina and silica
tetrahedra with the edges of the particles possessing typical
oxide characteristics and the surfaces exhibiting a negative
charge resulting from substitutions in the crystal lattice.
Cations that are adsorbed on the surface or contained in between
the sheets are exchangeable. Cations of the added salt will
exchange for the same reason with protons on the platelets and
the resultant increase in acidity will lead to dissolution of
aluminum ions. Readsorption of aluminum complexes and deposition
of aluminum by hydroxy colloids has been considered to be a major
factor responsible for slow changes in properties of kaolinite in
electrolyte solutions.[31,32] Evidently such slow changes in
surface properties in salt solutions can lead to continued uptake
of the polymer as observed in the present study. The total rate
of adsorption itself will vary with ionic strength depending
mainly on changes in the solvent power of the medium.

The rate of uptake of polymer in alkaline solutions is shown
in Figure 7. In this pH range, while the initial adsorption
kinetics are of a similar type in the presence and in the absence
of salt, the slow adsorption step is of a different type in the
absence of salt. Adsorption is in fact found to decrease under
this condition and this is attributed to the hydrolysis of the
polymer in the basic medium. Polyacrylamide, upon hydrolysis,
behaves like acrylamide-acrylic acid copolymer leading apparently
to some desorption of the polymer. Such desorption has actually
been observed by Martensen.[9] The above explanation is supported
by the observed increase in the viscosity of the PAM in alkaline
salt solutions (Figure 3).

It can be seen from the present results that the rate of adsorption of polymers on minerals can be a complicated function of the properties of the original adsorbent and adsorbate materials, as well as changes in these properties due to mineral dissolution, polymer hydrolysis and degradation, etc. Valuable information on the behavior of adsorbed polymers can be obtained from studying differences in rates of adsorption under various solution conditions.

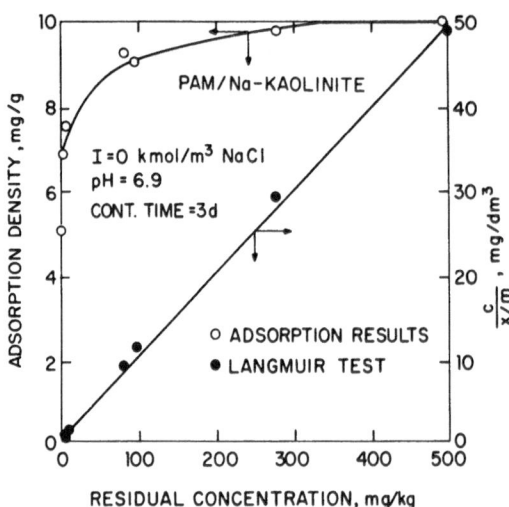

FIGURE 8   Typical adsorption isotherm of polyacrylamide on Na-kaolinite as a function of residual polymer concentration and the corresponding Langmuirian plot.

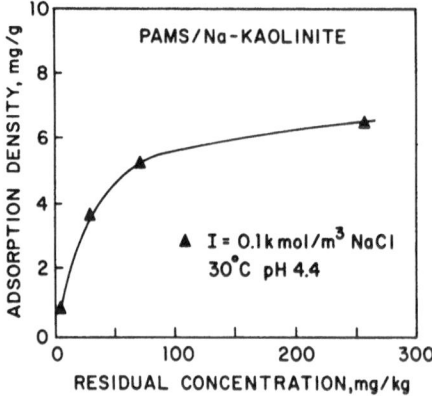

FIGURE 9   Adsorption isotherm of sulfonated polyacrylamide on Na-kaolinite as a function of residual polymer concentration.

Adsorption Isotherms

Based on the results obtained for the rate of increase of adsorption, an equilibration time of three days was chosen for the study of the dependence of adsorption on different system variables. It is evident from the kinetic data that adsorption is nearly complete within that time interval.

Typical adsorption isotherms obtained for the PAM and PAMS on Na-kaolinite are given in Figures 8 and 9. Isotherms under other pH and ionic strength conditions are discussed elsewhere.[30] The isotherms in Figures 8 and 9 have the characteristic features of those usually obtained for macromolecules. When only minor amounts of the polymer are introduced into the system, it is nearly totally removed from the solution by adsorption. Above a certain coverage, the adsorption of PAM is found to be Langmuirian in general as suggested by the Langmuir plot in Figure 8. Even though the fit to the Langmuir equation is good, the isotherm can be considered only as pseudo-Langmuirian since polymer adsorption cannot be expected to satisfy all the required criteria for Langmuir-type adsorption. For example, the adsorption of polymers is essentially irreversible and, furthermore, an equilibrium between the adsorbate and the bulk solution is not attained within the time intervals indicated. Useful information can be obtained, however, by assuming a Langmuir model within a reasonable concentration range. The slope of the plot in Figure 8, for example, is a measure of PAM adsorption at infinite concentration. Area per molecule under this condition was calculated to be 185 $\overset{\circ}{A}^2$. Assuming a segment size of 3 x 5 $\overset{\circ}{A}$, the maximum amount of polymer that can be adsorbed for flat coverage by this polymer is 0.5 mg/m^2. Since the maximum adsorption density obtained for the present system is 1.2 mg/m^2, about 40% of the segments can be considered to be in actual contact with the surface. This high segment contact suggests a relatively large negative free energy of adsorption for the PAM kaolinite.

The effect of pH and ionic strength on the adsorption is conveniently seen by replotting adsorption data as a function of these variables at constant residual polymer concentration.

Effect of pH

Adsorption dependence on pH is shown in Figure 10. At lower ionic strengths of 10^{-2} and 10^{-1} M, adsorption remains independent of pH from pH 2 to about pH 7 and decreases above pH 7. At higher ionic strengths, however, adsorption is found to decrease continuously in the complete pH range. The decrease in adsorption at high pH values can be attributed to increased electrostatic repulsion between the negatively charged kaolinite particles and

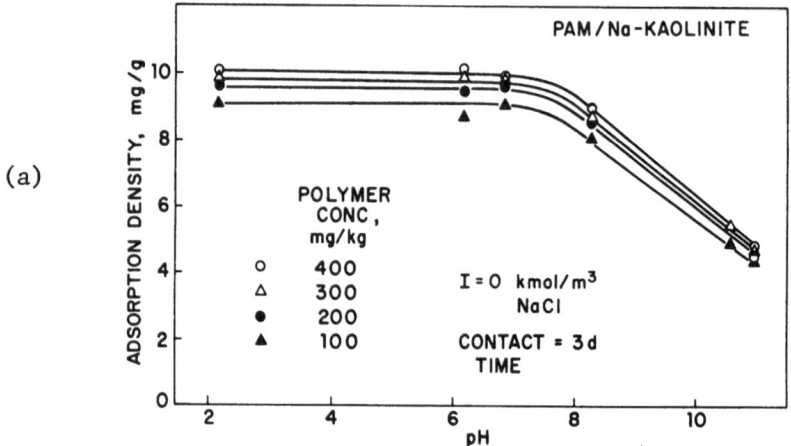

(a)

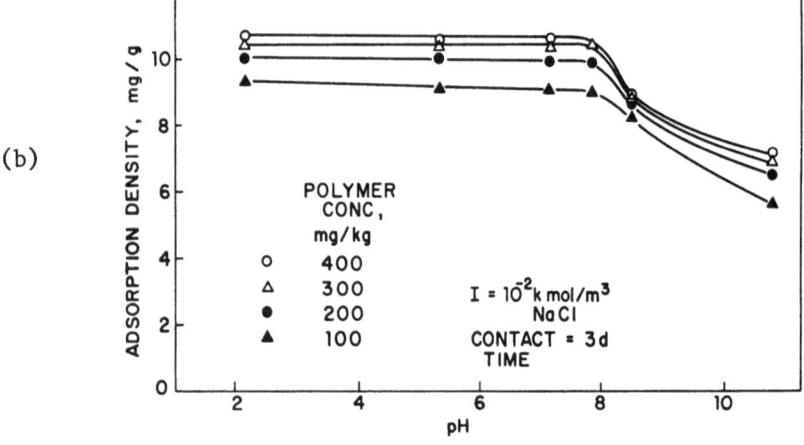

(b)

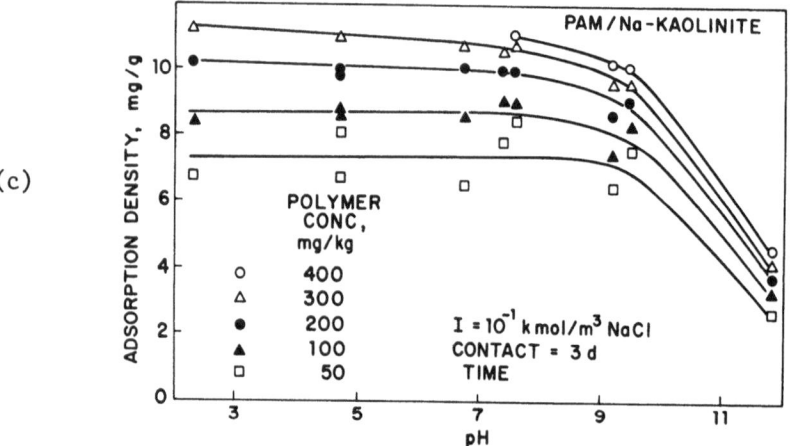

(c)

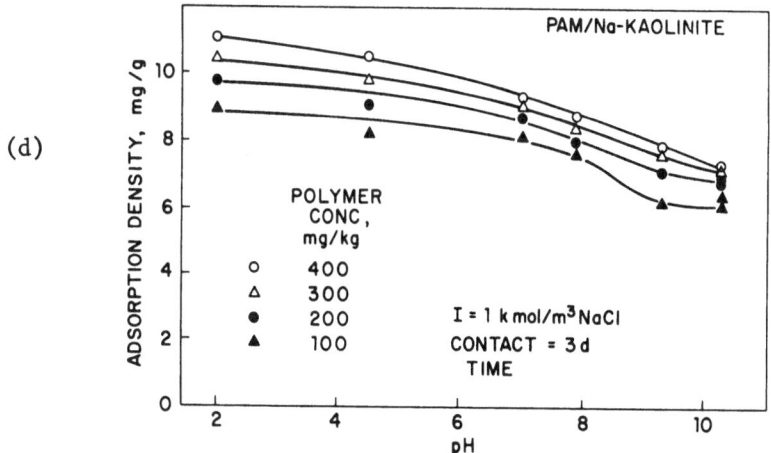

(d)

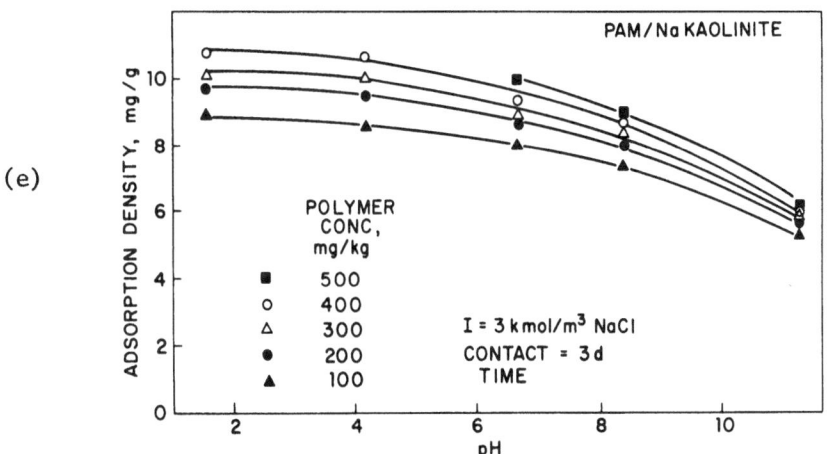

(e)

FIGURE 10   Adsorption of polyacrylamide on Na-kaolinite as
            a function of pH at   (a) in the absence of
            supporting electrolyte,   (b) in 10^{-2} M NaCl,
            (c) in 10^{-1} M NaCl,   (d) in 1 M NaCl, and
            (e) in 3 M NaCl solutions.

PAM molecules that are partially ionized in alkaline solutions.
Indeed such an effect should have been minimal under high ionic
strength conditions.   The reasons for the observed dependency of
adsorption on pH in 1 M and 3 M solutions are not known at this
time.   In addition to electrostatic adsorption it appears that
the possible effect of charging of the molecules on uncoiling of
the polymer chain and the resultant changes in entropy at higher
pH values might be playing a major role.   Also the degree of
hydrogen bonding, the major mechanism that has been considered in

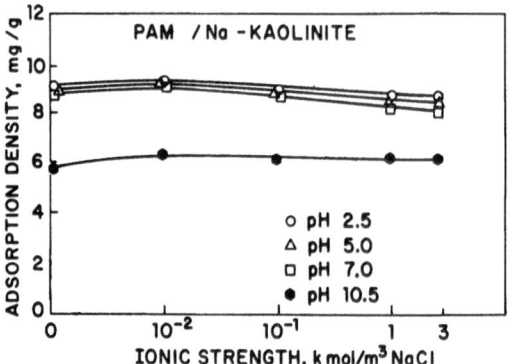

FIGURE 11   Adsorption of polyacrylamide on Na-kaolinite as a
function of ionic strength.

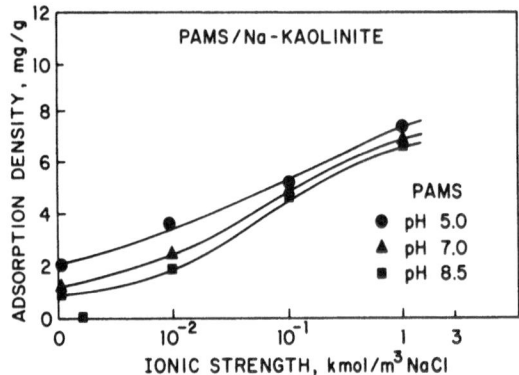

FIGURE 12   Adsorption of sulfonated polyacrylamide on Na-
kaolinite as a function of ionic strength.

the past to be responsible for the adsorption of these polymers,
might decrease with increase in pH owing to a decrease in the
number of silanol and aluminol groups on the clays as the surface
sites ionize to a greater degree under those conditions.

Effect of Ionic Strength

The effect of ionic strength on the adsorption of PAM and
PAMS is shown in Figures 11 and 12. While there is a significant
effect of ionic strength on the adsorption of PAMS in general,
there is first a slight increase in the adsorption of PAM and
then a measurable decrease with increase in the ionic strength.
Ionic strength can be expected to affect adsorption due to its

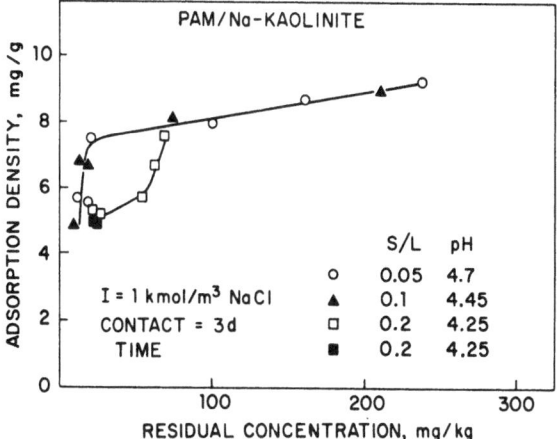

FIGURE 13   Adsorption of polyacrylamide on Na-kaolinite at
             different solid-to-liquid ratios.

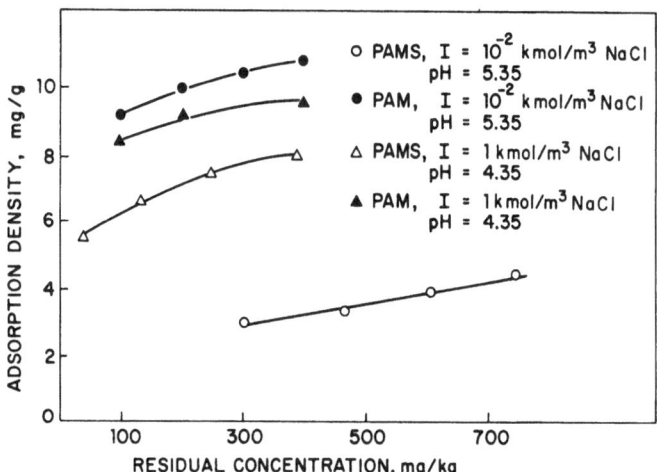

FIGURE 14   Comparison of the adsorption dependence of poly-
             acrylamide on the ionic strength of the solution
             to that of sulfonated polyacrylamide.

influence on the solvent power as well as on electrostatic and
hydrogen bonding forces. While the influence of salt on solvent
power will be to increase adsorption, electrostatic adsorption of
PAMS or the hydrolyzed PAM on positive sites (such as that of
aluminum) will be reduced by it. On the other hand, any electrostatic repulsion between the PAM or PAMS and the negative clay
surface will be reduced by salt addition leading to an increase
in the adsorption. The cumulative result of any change in ionic
strength will be determined by the relative magnitude of all the
above effects.

Effect of Solid-to-Liquid Ratio

     The effect of solid-to-liquid ratio is illustrated in Figure 13.
Ideally, changes in the solid-to-liquid ratio are not expected to
have any influence on the equilibrium adsorption density unless
there are major interactions between the particles themselves.
In the present case, there is no measurable effect of an increase
in the solid-to-liquid ratio from 0.05 to 0.1, but further increase
to 0.2 is found to produce a decrease in adsorption density at
the low polymer concentrations investigated. We attribute this
to the larger segment to surface contact that is possible under
such conditions. With increase in polymer concentrations as the
saturation adsorption density is achieved, the polymer is squeezed
out to form loops and it will then adsorb at all solid/liquid
ratios probably to a maximum density dictated entirely by the
physico-chemical conditions of the system (equilibrium polymer
concentration, pH, ionic strength, and temperature).

Polymer Charge

     The role of the charge on the polymer is illustrated in
Figure 14 where the adsorption of the anionic PAMS is compared
with that of PAM. As expected, the anionic PAMS is found to
adsorb less than the relatively uncharged PAM. In 10^{-2} molar
salt solution the adsorption is almost one-fifth when the sulfonate
groups are incorporated into the polymer. At high ionic strengths,
the electrostatic repulsion between the anionic sulfonate groups
and the negative surface sites on kaolinite is minimized and the
adsorption of PAMS is close to that of PAM. It is interesting
that while the increase in ionic strength as mentioned earlier
reduces the adsorption of PAM only slightly, it produces a significant increase in the adsorption of the sulfonated polyacrylamide.

Effects of Temperature

     The adsorption data for PAM and PAMS at two different tem-
peratures are given in Figure 15. The increase in temperature is
found to produce a slight decrease in adsorption of PAM and no
decrease in that of PAMS. Like the ionic strength effects,
temperature effects can also be rather complex due to the simultaneous influence it has on a number of system properties such as
solvent power, mineral solubility, electrostatic adsorption,
hydrogen bonding and polymer configuration. Data for adsorption
under a wider range of conditions are needed to identify major
factors responsible for the temperature effects on polymer adsorption.

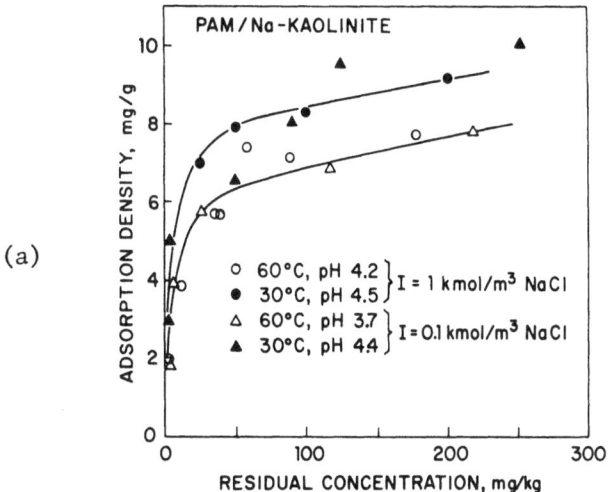

(a)

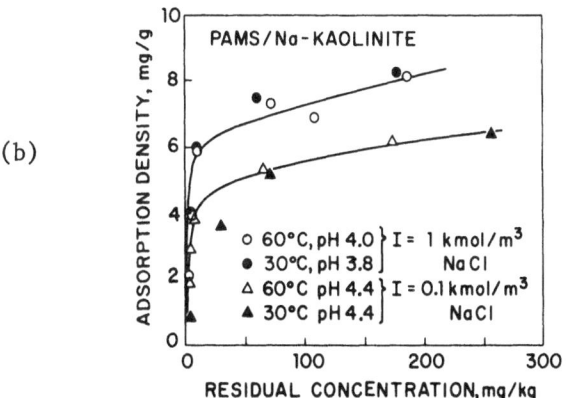

(b)

FIGURE 15   Adsorption of (a) polyacrylamide and (b) sulfonated
             polyacrylamide on Na-kaolinite at two different
             temperatures.

SUMMARY AND CONCLUSIONS

1.   Adsorption of polyacrylamide and sulfonated polyacrylamide
     was studied as a function of pH, ionic strength, temperature
     and solid-to-liquid ratio. Polymers were characterized
     using intrinsic viscosity measurements. The ionic character
     of PAMS was confirmed by its higher intrinsic viscosity and
     the larger effect of ionic strength and pH on its viscosity
     than that on the viscosity of PAM. Under alkaline and high
     temperature conditions, PAM exhibited viscosity changes that
     are indicative of a tendency to hydrolyze. The viscosity
     alterations were suppressed in the presence of salt.

2.   Kinetics of adsorption included a fast and a slow step, the
      slow step being a pseudo-Langmuirian adsorption. The rate
      of initial rapid adsorption was reduced markedly by salt
      additions owing to a constricted polymer configuration
      and/or increased acidity of kaolinite in salt solutions.
      While the initial adsorption might be mostly diffusion-
      controlled, the subsequent slow increase in adsorption is
      considered to be due to the changes in surface properties of
      kaolinite owing to dissolution and re-adsorption of colloids
      of aluminum complexes and possibly the changes in the con-
      figuration of the adsorbed polymer. Some desorption was
      observed in alkaline solutions under prolonged conditioning,
      possibly due to the hydrolysis of the polymer.

3.   Introduction of anionic sulfonate groups into the polymer
      markedly reduced adsorption due to increased electrostatic
      repulsion between the polymer and the kaolinite particles.

4.   Increase in pH caused a decrease in adsorption of PAM in the
      alkaline range at low ionic strengths (further suggesting
      some hydrolysis) and in the complete pH range at higher
      ionic strengths. Similar increases in pH caused, as expected,
      a decrease in adsorption of PAMS under all conditons. The
      pH effect is attributed mainly to the electrostatic repulsion
      between PAMS and kaolinite particles that become increasingly
      negatively charged with increase in pH.

5.   Ionic strength increase, below 10^{-2} M, caused a slight
      increase in adsorption of PAM and above 10^{-2} M a decrease.
      In the case of PAMS, adsorption increased with increase in
      ionic strength under all conditions.

6.   While changes in temperature had no significant effect on
      the adsorption of PAMS, PAM adsorption was reduced by an
      increase in temperature.

7.   Solid-to-liquid ratio showed an effect at the highest ratio
      studied, i.e., reduced adsorption of PAM at this ratio.

8.   The above effects are the cumulative result of various
      changes in the configuration of the polymer (adsorbed and
      dissolved) surface properties of kaolinite particles owing
      to processes involving dissolution and readsorption of
      various dissolved species, particularly of alumina. The
      role of hydrogen bonding and electrostatic bonding in deter-
      mining the adsorption of PAM and PAMS under various conditions
      is considered.

ACKNOWLEDGEMENTS

   The authors acknowledge the support of the Thermodynamics
and Mass Transfer Program of the National Science Foundation
(ENG-78-11776), Amoco Production Company, Chevron Oil Field
Research Company, Exxon Research and Engineering Co., Gulf Research
and Development Company, Marathon Oil Company, Mobil Research and
Development Company, Shell Development Company, Texaco Inc., and
Union Oil Company of California.

REFERENCES

1. Yu.S. Lipatov and L.M. Sergeeva, "Adsorption of Polymers",
    John Wiley & Sons, New York (1972).
2. T. Sato and R. Ruch, Adsorption of polymers from solutions,
    in: "Stabilization of Colloidal Dispersions by Polymer
    Adsorption", Dekker, New York (1980).
3. T.F. Tadros, J. Colloid Interface Sci. 64:36 (1978).
4. G.J. Fleer, "Polymer Adsorption and Its Effect on Colloidal
    Stability", Thesis, Meded. Landbouwhogeschool, Wageningen,
    71-20:12 (1971).
5. L.K. Koopla and J. Lyklema, Faraday Discuss. of Chem. Soc.
    59:230 (1975).
6. P. Somasundaran, J. Colloid Interface Sci. 31:557 (1969).
7. L. Jankovics, J. Polym. Sci. A.3:3519 (1965).
8. M.P. Nedelcheva and G.V. Stoilkow, Colloid and Polym. Sci.
    255:327 (1977).
9. J.L. Martensen, Proc. Natl. Conf. Clays and Clay Minerals
    9:530 (1961).
10. A.P. Black, F.B. Birkner and J.J. Morgan, J. Colloid Interface
    Sci. 21:626 (1966).
11. K. Roberts, J. Kowalewska and S. Friberg, J. Colloid Interface
    Sci. 48:363 (1974).
12. A.S. Michaels and O. Morelos, Ind. & Eng. Chem. 47:1801
    (1955).
13. N. Schamp and J. Huylebroeck, J. Polymer Sci. Pt. C, Symp.
    No. 42:553 (1973).
14. B.P. Gyani, J. Indian Chem. Soc. 21:79 (1944).
15. T. Sato, J. Appl. Polym. Sci. 15:1053 (1971).
16. P. Kennedy, M. Petronio and H. Gisser, J. Phys. Chem.
    75:1975 (1971).
17. W. Heller, Pure Appl. Chem. 12:249 (1966).
18. H.L. Frisch, M.Y. Hellman and J.L. Lundberg, J. Polym. Sci.
    38:441 (1959).
19. T. Sato, T. Tanaka and T. Yoshida, J. Polym. Sci., Polym.
    Lett. 5:947 (1967).
20. B.J. Fontana and J.R. Thomas, J. Phys. Chem. 65:480 (1961).

21. J. Koral, R. Ullmann and F.R. Eirich, J. Phys. Chem. 62:541 (1958).
22. R. Simha, H.L. Frisch and F.R. Eirich, J. Phys. Chem. 57:584 (1953).
23. H.L. Frisch and R. Simha, J. Phys. Chem. 58:507 (1954).
24. H.L. Frisch, J. Phys. Chem. 59:633 (1959).
25. A. Silberberg, J. Phys. Chem. 66:1872 (1962).
26. A. Silberberg, J. Phys. Chem. 66:1884 (1962).
27. A. Silberberg, J. Phys. Chem. 48:2835 (1968).
28. A. Silberberg, J. Phys. Chem. 46:1105 (1967).
29. T. Wada, H. Sekiya and S. Machi, J. Appl. Polym. Sci. 20:3233 (1976).
30. A.F. Hollander, P. Somasundaran and C.C. Gryte, paper submitted for publication.
31. H.S. Hanna and P. Somasundaran, J. Colloid Interface Sci. 70:181 (1979).
32. P. Somasundaran and H.S. Hanna, Soc. Petrol. Eng. J. 19:227 (1979).

INTERACTION OF PESTICIDES WITH CHITOSAN

Pam Davar and James P. Wightman

Chemistry Department
Virginia Polytechnic Institute and State University
Blacksburg, Virginia 24061

ABSTRACT

The uptake isotherms of 2,4-D, Dicamba, 2,4,5-T, MH and MCPA
have been measured on chitosan at room temperature. Chitosan is
the deacetylated derivative of the natural marine polymer, chitin.
Time of equilibration was determined for each system and the
effects of solution pH on pesticide uptake have been studied.
The adsorption of 2,4-D on activated charcoal was measured for
comparison. Ancillary measurements of the chitosan samples after
pesticide uptake included SEM/EDAX (scanning electron microscopy/
energy dispersive analysis of X-rays) and ESCA (electron spectroscopy for chemical analysis).

INTRODUCTION

Chitin (poly-N-acetyl D-glucosamine) discovered by Henri
Braconnot in 1811 is probably the second most abundant polysacchar-
ide, forming the exoskeleton of most of the invertebrates.[1] The
chitin content of dry shellfish waste varies between 14-35%.[2]
The U.S. commercial landings of crab and shrimp in 1978 amounted
to 872 million pounds.[3] Thus, large quantities of natural chitin
are produced during the processing of crabs and shrimp. Chitosan
is the deacetylated form of chitin. An increasing awareness that
chitin/chitosan have an impressive number of potential applications
in both basic and applied science and technology has stimulated
new research which was recently summarized.[4]

The sorption properties of chitin/chitosan were first studied
by Hackman[5] in relation to the sorption of proteins from aqueous

163

solution. This was followed by Giles et al.[6,7] who reported
observations on chitin as a sorbent for organic and inorganic
acids. In the past several years, chitin and chitosan have been
used as sorbents for thin layer chromatography,[8,9] as a chromatographic support for high pressure liquid chromatography,[10] and
as a chelating polymer for bonding toxic metal ions.[11-13]

In spite of the natural abundance of chitin and the widespread
use of pesticides, it is perhaps surprising that very little
research has been conducted on the interaction of pesticides with
the polymer. Richards and Cutkomp[14] showed that chitin binds
DDT. Following this, Lord[15] determined that chitin as a colloidal
suspension sorbs DDT. He established that DDT sorption was a
surface phenomenon and that it occurred by a physico-chemical
process associated with the presence of amino groups in the
chitin.

Kemp and Wightman[16] reported preliminary results on the
uptake of 2,4-D and Dicamba by chitin and chitosan. McCormick et
al.[17] studied the effectiveness of chitosan as a polymeric matrix
for controlled release pesticide systems. Weber[18] categorized a
large number of organic pesticides as acidic, basic, cationic and
non-ionic according to their significant chemical properties.

The objective of this work was to determine the type of
pesticides capable of interacting with chitosan by measurement of
the uptake isotherm and to establish the uptake mechanism using
ancillary spectroscopic techniques.

EXPERIMENTAL

Materials

Chitosan was obtained from the Velsicol Chemical Corporation,
Chicago. A number of pesticides, representative of all four
groups, were tested initially. These were obtained from the
Environmental Protection Agency, Research Triangle Park, N.C. and
were of technical grade.

Negligible sorption of Atraton and Prometon (basic pesticides),
Dimethoate and Propham (non-ionic pesticides), and, Diquat dibromide
(cationic pesticide) was noted on chitosan. Sorption studies
were continued with 2,4-D, 2,4,5-T, Dicamba, MH, and MCPA since
only acidic pesticides showed significant uptake. The chemical
name, structure, molecular weight, characteristic wavelength of
maximum absorption (λ_{max}), solubility, supplier, and analytical
purity for each pesticide are listed in Table 1.

Activated charcoal (6-14 mesh) in granular form was obtained
from Matheson, Coleman and Bell.

Table 1. Description of pesticides.

Pesticide	Chemical Name	Structure	Mol. Wt.	λ_{max}(Å)	Solubility (ppm)	Supplier	Analytical purity
2,4-D	2,4-dichloro phenoxyacetic acid		221	2815(Molecular) 2833(Anionic)	650	National Bio-chemical Corp.	100%
2,4,5-T	2,4,5-trichloro phenoxyacetic acid		255.5	2870	238	Diamond Chemicals	99+%
Dicamba	2-methoxy-3,6 dichloro benzoic acid		221	2800	4500	Velsicol Chemical Corp.	87.7%
MCPA	(4-chloro-2-methylphenoxy acetic acid		201	2780	500	Union Carbide	95%
MH	6-hydroxy-3-(2H)-pyrid-azinone		112	3020	6000	Uniroyal Inc.	99%

Procedures

All sorption studies were done by equilibrating 0.1 g chitosan
(or 50 mg charcoal) with 50 mL of aqueous pesticide solutions of
known concentration. Calibration plots of pesticide concentrations
vs. absorbance in the UV range were made for each pesticide
system, using a Cary-14 spectrophotometer. All absorbance values
were corrected for the reference, which was either water or water
equilibrated with chitosan. Pesticide concentrations before and
after equilibration were obtained by interpolation of these
plots. Sorption was measured as the product of the change in
pesticide concentration and the solution volume normalized by the
amount of chitosan used.

Nitric acid concentration was obtained directly from a
measure of solution pH. All pH determinations were done with an
Orion Research digital pH meter and a Thomas combination electrode.
Experiments were carried out at room temperature unless otherwise
specified.

The time of equilibration was determined for 10^{-3} M solutions
of 2,4-D, 2,4,5-T, Dicamba and HNO_3 on chitosan and for 2,4-D on
charcoal. The change of pH with time was determined by equilibrating a 9 x 10^{-4} M solution of 2,4-D with chitosan. To investigate the effect of increasing solution pH on sorption, 2,4-D
solutions were prepared in 0.1 M NaH_2PO_4/Na_2HPO_4 buffer with pH
values varying from 4.56 to 6.50. For the 2,4-D solution, the pH
was also adjusted by addition of 0.1 M HNO_3 or 10^{-6} M NaOH. The
pH of these solutions were held constant between 4.5-8.0.

Equilibrated solutions were filtered to recover the chitosan,
which was then air-dried for about 30 mins. Dried chitosan
powder was mounted on double stick tape for analysis by ESCA
(electron spectroscopy for chemical analysis) and on copper
conductive tape for SEM/XRS (scanning electron microscopy/X-ray
spectroscopy) analysis. Photomicrographs of chitosan before and
after equilibration were obtained, using an Advanced Metals
Research 900 scanning electron microscope. Qualitative analysis
of the sorbent was done by an EDAX (energy dispersive analysis of
X-rays) attachment on the microscope. ESCA spectra of the
chitosan samples were obtained using a DuPont 650 electron spectrometer with a Mg X-ray source. The Cls photopeak at 284.6 eV was
used as a reference for binding energy calibration.[19] Curve
fitting of the Nls photopeak was done for a chitosan sample
equilibrated with 2,4,5-T.

RESULTS AND DISCUSSION

Sorption Isotherms

Significant uptake was realized for all acidic pesticides on chitosan, as shown by the isotherms in Figures 1 and 2. The isotherms are similar to the S1 type as described by Giles et al.[20] The initial curvature of the isotherm suggests a side-by-side association between the adsorbed molecules, helping to hold them to the surface. This has been called "co-operative adsorption", with the solute molecules tending to be adsorbed, packed in rows or clusters.[20,21] The steep slope has been attributed to multilayer adsorption at high relative concentrations.[21,22] Care has to be exercised, however, in this particular interpretation of the isotherm shape since, as is pointed out below, sorption on chitosan is a complex process and may not be amenable to such interpretation by a simpler adsorption process.

Isotherms of similar type have been reported for Dasanit on montmorillonite clay, saturated with various cations[23] and for protein adsorption on chitin.[5] On the other hand, Dole[24] reported linear isotherms suggesting a solution process for the uptake of nitrophenol on Nylon from aqueous solutions.

The active sites in chitosan are believed to be the amino groups. The pesticides are organic acids with pK_a values of approximately 2.0-4.0.[18] In aqueous solution these acids are dissociated according to Equation (1).

$$RCOOH \rightleftharpoons RCOO^- + H^+ \tag{1}$$

$$or \quad ROH \rightleftharpoons RO^- + H^+$$

The amine sites in chitosan may be protonated according to Equation (2).

$$R'-NH_2 + H^+ \rightleftharpoons R'NH_3^+ \tag{2}$$

Sorption can then take place through an electrostatic interaction of the carboxyl anion and the protonated amine site as indicated below

$$RCOC^\ominus + R'NH_3^\oplus \rightleftharpoons RCOO^\ominus \quad {}^\oplus H_3NR' \tag{3}$$

The uptake process involves coupled equilibria which are pH dependent. The hydroxyl groups in the polymer are believed to be strongly solvated in water, thus minimizing sorption of pesticides on those sites.

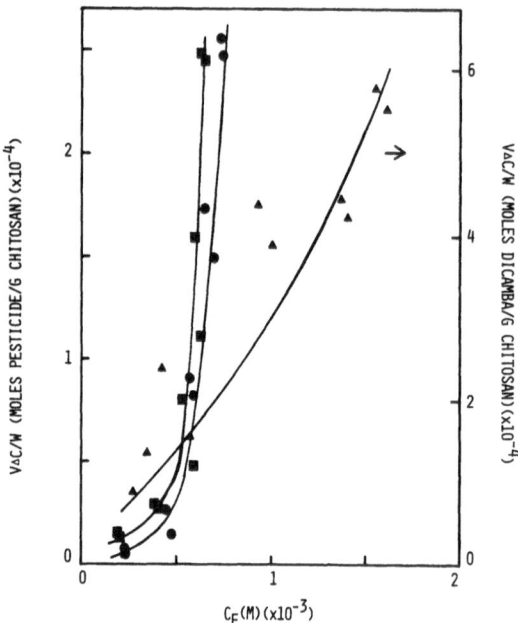

FIGURE 1:   Sorption isotherms for MCPA (●), Dicamba (▲) and
            2,4-D (■) on chitosan.

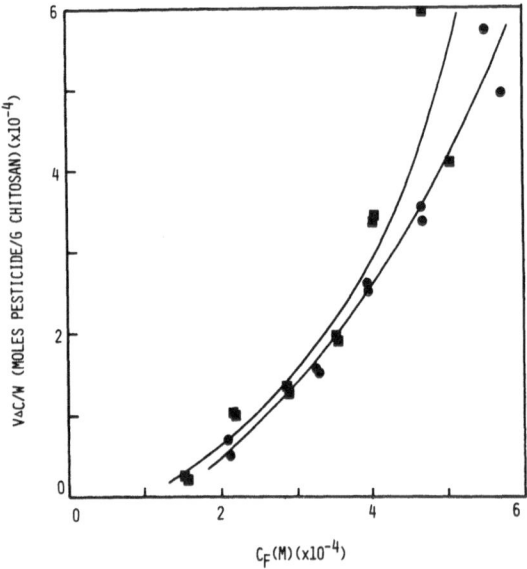

FIGURE 2:   Sorption isotherms for MH (●) and 2,4,5-T (■) on
            chitosan.

In order to compare the sorption of 2,4-D on chitosan to a
less complex solid, sorption was measured on charcoal. The
results are shown in Figure 3 for two different weights of charcoal.
Sorption was independent of sample weight. The sorption of 2,4-D
on charcoal is an L2 type as described by Giles.[20] This isotherm
shape was also noted by Dole[24] for nitrobenzene adsorption on
Graphon from aqueous solution. The initial curvature shows that
as more sites in the substrate are filled, it becomes increasingly
difficult for a solute molecule to adsorb on a vacant site. This
is indicative of "monolayer adsorption" of molecules, with very
little, if any, competition from the solvent. In contrast to the
sorption process involved with chitosan, adsorption on the charcoal
is basically a "pore-filling" process, in which the available
volume of the pores is the controlling factor.

Time of Equilibration

An interesting trend was noted in the relationship between
the amount of each pesticide sorbed on chitosan and time elapsed
as illustrated for 2,4-D in Figure 4. A sharp increase and then
a steady decline in the amount sorbed was seen, before the system
reached steady state. Times of equilibration with chitosan
varied from 24 hrs for 2,4-D and Dicamba, to 44 hrs for 2,4,5-T.
The equilibration time for 2,4-D on charcoal was 55 hours. This

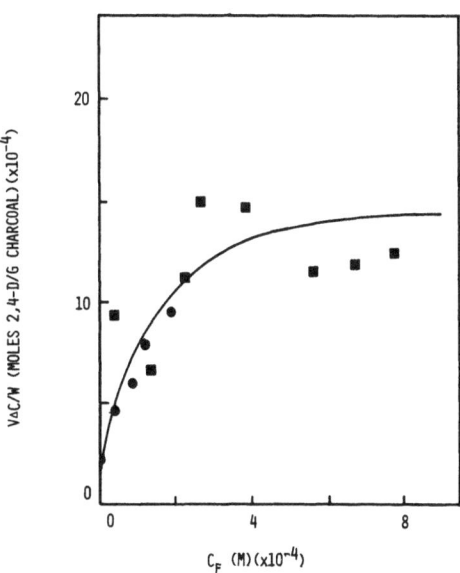

FIGURE 3:   Sorption isotherms for 2,4-D on 10 mg (■) and 50 mg
             (●) on charcoal.

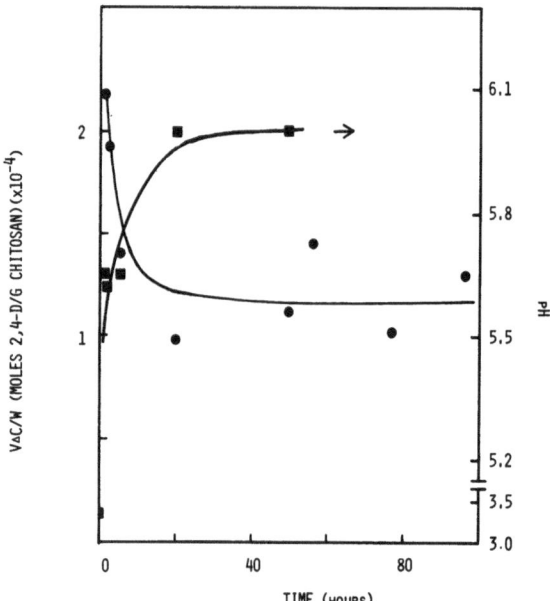

FIGURE 4:   Uptake and pH as a function of time for 2,4-D (initial
            concentration: 9 x 10^{-4} M) on chitosan.

long equilibration time can be attributed to slow diffusion into
micropores known to be present in charcoal.[25]

     Significantly, as the pesticide sorption by chitosan decreased
with time, there was a concomitant increase in the pH of the
solution as shown also in Figure 4. Note that the time to reach
steady state at a pH of 6.0 was about 24 hours for both the
sorption and pH results. This peculiar dependence of uptake with
time suggests a complex sorption process.[23]

pH

     Change in solution pH has a significant effect on 2,4-D
sorption in chitosan as shown in Figure 5. Maximum sorption
occurred between pH 4.5 and 5.0. Distribution of various species
present in solution at a particular pH may be calculated according
to Equations (4) and (5):

$$pH = pK_a - \log \frac{[HA]}{[A^-]} \qquad (4)$$

$$C_T = [HA] + [A^-] \qquad (5)$$

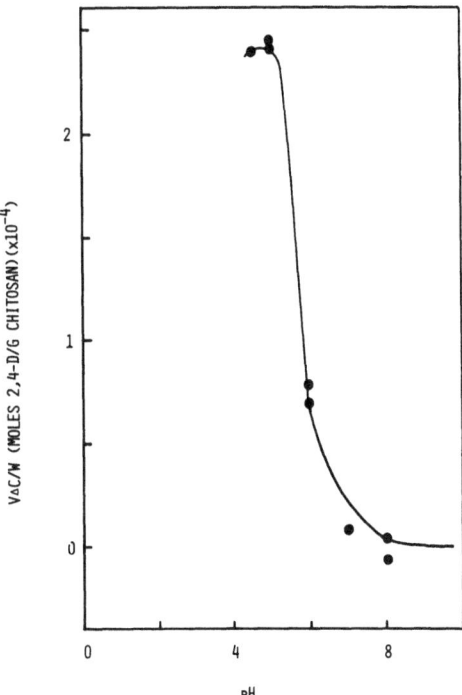

FIGURE 5:   Effect of pH on the uptake of 2,4-D (initial concentra-
            tion: 9 x 10^{-4} M) on chitosan.

where C_T refers to total initial concentration. [HA] is equal to
the concentration of the acid form of a species and [A⁻] is equal
to the concentration of the base form of a species. Using these
equations and pK_a values of 6.3 and 2.8 for chitosan[10] and 2,4-D,
respectively, concentrations of $R'NH_3$, $R'NH_2$ and the acid and
base forms of 2,4-D were calculated.[26] The conclusion drawn from
the calculations is that electrostatic effects dominate between
the pK_a values of the two reacting species. Maximum sorption
would therefore be expected to occur between pH 3 and 6 consistent
with the experimental observation. Similar results have been
reported by Giles et al.[7] for systems of azo compounds sorbed on
chitin in which case maximum adsorption occurred at pH 4.7.

    No significant uptake of 2,4-D on chitosan was noted when
the solution pH was adjusted with sodium phosphate buffer. This
may be due to preferential adsorption of the added ions, which
are in relatively high concentration. The effects of cations and
anions on sorption by chitin have been reported.[27,28]

SEM/EDAX

    Representative SEM photomicrographs in Figure 6 of chitosan
samples before and after equilibration with 2,4-D show no dis-

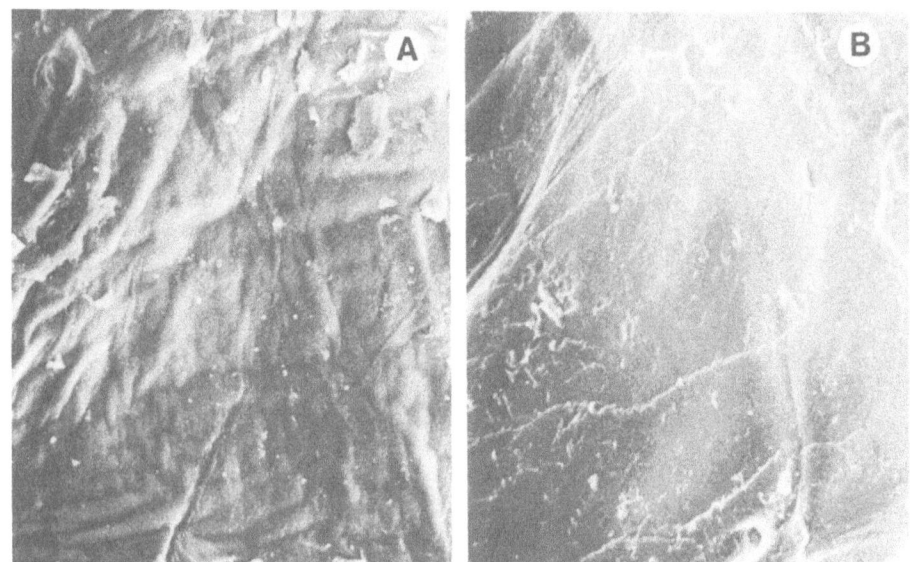

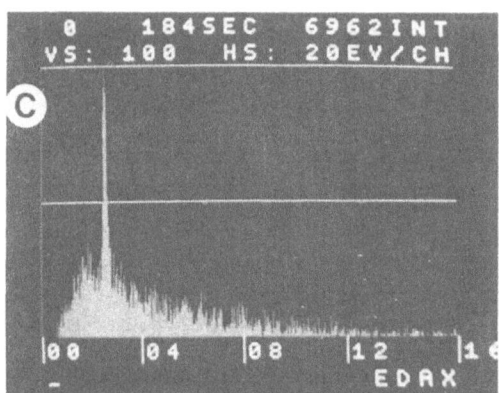

FIGURE 6:   SEM photomicrographs of chitosan before (A) and after
            (B) equilibration with 250 ppm 2,4-D showing EDAX
            spectrum (C).

tinctive differences in surface features of the polymer. This
same result was observed for the other acidic pesticides. Sorption
of the pesticides, therefore, does not result in changes in the
surface morphology of chitosan. A significant chlorine signal
was seen in the EDAX spectrum of the chitosan after equilibration
with 2,4-D as illustrated also in Figure 6. Again, this was a
general result observed also for the other chlorinated pesticides.

This EDAX result demonstrates, independently, the uptake of the pesticides by chitosan supporting the measured isotherms. The observed chlorine signal, perhaps more importantly, suggests a sorption mechanism rather than adsorption. Kang, Skiles and Wightman[29] have shown that if chlorine species are restricted on the surface of a solid (i.e., adsorption), no chlorine EDAX signal is observed. Thus, there is evidence that the pesticides diffuse into the chitosan matrix (sorption) rather than being localized on the chitosan surface (adsorption).

ESCA

        C 1s, O 1s, N 1s and Cl 2p photopeaks for chitosan equili-brated with 2,4-D are shown in Figure 7. Signal intensities were corrected with the photo-electron cross sections for each element.[30] A significant Cl 2p peak for the equilibrated compared to the unequilibrated chitosan samples again demonstrates the uptake of the chlorinated pesticides by chitosan.

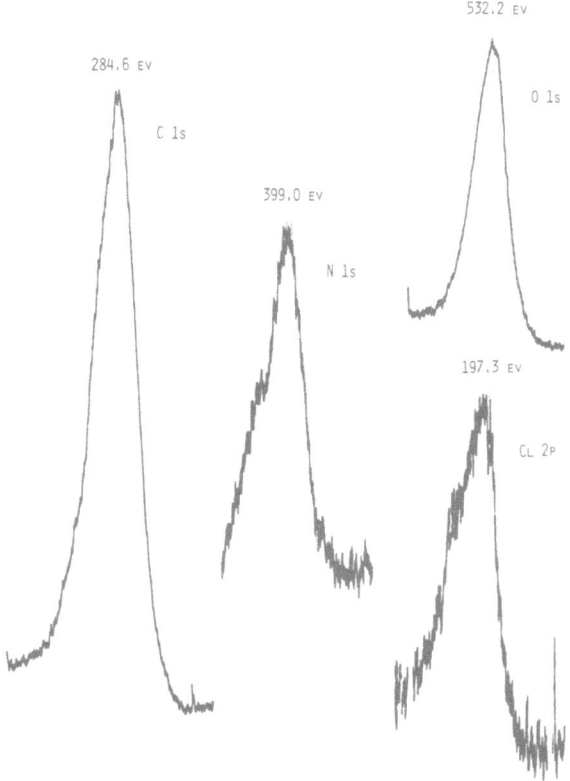

FIGURE 7:   ESCA photopeaks of chitosan after equilibration with
             250 ppm 2,4-D.

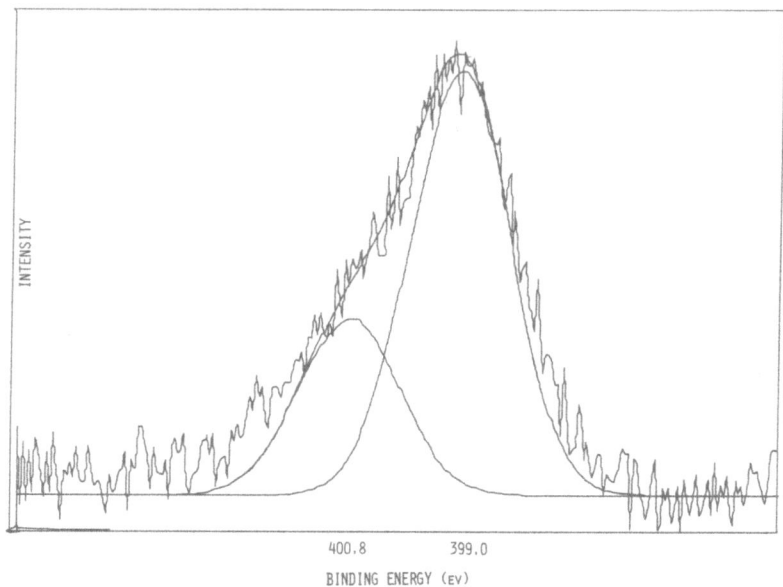

BINDING ENERGY (EV)

FIGURE 8:   Curve fit spectrum for Nls photopeak of chitosan after
            equilibration with 120 ppm Dicamba.

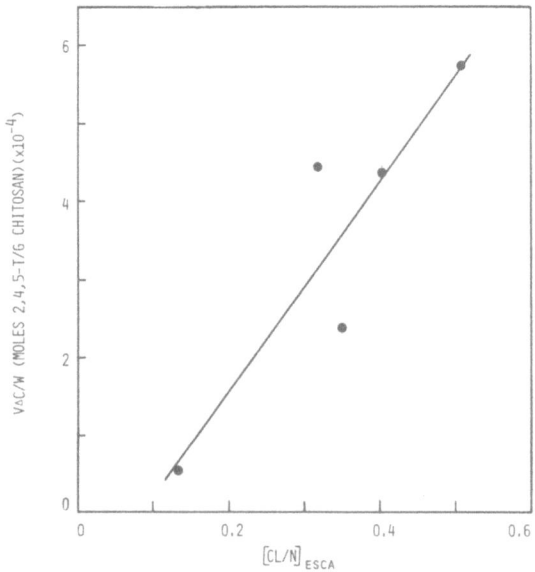

FIGURE 9:   Comparison of the uptake of 2,4,5-T on chitosan
            determined from the isotherms with the Cl/N ratios
            determined from ESCA photopeak intensities.

The N 1s photopeak included a high binding energy shoulder where the intensity appeared to increase with increasing pesticide concentration. The composite peak for the equilibrated chitosan was curve fitted, therefore, into two separate peaks. One peak occurred at a binding energy of 399.0 eV and the other at 400.8 eV shown in Figure 8. The former photopeak is assigned to nitrogen in the unreacted amine sites in chitosan; the higher binding energy photopeak is assigned to nitrogen in amine sites interacting with the pesticides. The electron density around nitrogen in a reacted amine site would be expected to be less than in an unreacted site and hence give rise to a higher binding photopeak.

The above assignments are consistent with ESCA results for similar bonding reported in the literature. The nitrogen 1s photopeaks[31] in a primary amine and the hydrogen chloride salt of a primary amine occur at 398.7 and 401.0 eV. respectively. This result demonstrates the utility of the ESCA technique in defining the interaction site of pesticides with chitosan.

A comparison of the uptake isotherm and ESCA results is shown in Figure 9 for the Dicamba/chitosan systems. Here, Cl/N ratios determined from measured ESCA intensities parallel an increase in pesticide uptake from the measured isotherms.

CONCLUSIONS

Significant uptake was realized only for the acidic pesticides 2,4-D, 2,4,5-T, Dicamba, MCPA and MH on chitosan. An initial rapid uptake of 2,4-D on chitosan was followed by a decrease in the uptake amount until steady state was reached after 24 hours. Changes in solution pH contributed significantly to the extent of sorption, maximum uptake occurring between pH 4.5-5.0. A signifi- cant chlorine signal was observed with both SEM/EDAX and ESCA for chitosan equilibrated with the chlorinated pesticides. The N 1s photopeaks at 399.0 and 400.8 eV were assigned to reacted (with the pesticide) and unreacted amine sites in chitosan.

ACKNOWLEDGEMENTS

The authors thank Dr. S.W. Bingham and Dr. R.W. Young for their help in obtaining the pesticides.

REFERENCES

1. E.R. Pariser, and S. Bock, "Chitin and Chitin Derivatives,"
    MIT Sea Grant Publication 73-2, Cambridge, Mass. (1972).
2. G.G. Allan, J.R. Fos and N. Kong, in "Proc. First. Intl.
    Conf. on Chitin/Chitosan," Muzzarelli, R.A.A. and E.R. Pariser,
    Eds., MIT Sea Grant Publication 78-7, p. 82, Cambridge Mass.
    (1977).
3. Fisheries of the United States, 1978. Current Fishing
    Statistics No. 7800. National Marine Fisheries Service,
    National Oceanic and Atmospheric Administration, U.S.
    Department of Commerce.
4. "Proc. First Intl. Conf. on Chitin/Chitosan," Muzzarelli,
    R.A.A. and E.R. Pariser, Eds., MIT Sea Grant Publication
    78-7, Cambridge, Mass. (1977).
5. R.J. Hackman, Austr. J. Chem. 8:530 (1955).
6. C.H. Giles, A.S.A. Hassan, M. Laidlaw, and R.V.R. Subramanian,
    J. Soc. Dyers Colourists 74:647 (1958).
7. C.H. Giles, and A.S.A. Hassan, ibid 74:682 (1958).
8. L. Lepri, P.D. Desideri, and R.A.A. Muzzarelli,
    J. Chromatogr. 139:337 (1977).
9. M. Takeda, in "Proc. First. Intl. Conf. on Chitin/Chitosan,"
    R.A.A. Muzzarelli, and E.R. Pariser, Eds., MIT Sea Grant
    Publication 78-7, pp. 355-363, Cambridge, Mass. (1977).
10. R.A.A. Muzzarelli, "Chitin," Chap. 5, Pergamon Press, London
    (1977).
11. R.A.A. Muzzarelli, Talanta 16:1571 (1969).
12. T-C. Hung and S.L. Han, Acta Oceanogr. Taiwan 7:56 (1977).
    Chem. Abstr. 89:8073t.
13. R.A.A. Muzzarelli, "Natural Chelating Polymers," Chap. 5,
    Pergamon Press, London (1973).
14. A.G. Richards and L. Cutkomp, Biol. Bull 90 (1946).
15. K.A. Lord, Biochem. J. 43:72 (1948).
16. M.V. Kemp and J.P. Wightman, Va. J. Science, in press.
17. C.L. McCormick and D.K. Lichatourich, J. Poly. Sci. Polym.
    Lett. Ed. 17:479 (1979).
18. J.B. Weber, in Adv. Chem. Series. No. 111. R.F. Gould, Ed.,
    pp. 55-120, Am. Chem. Soc., Washington (1972).
19. P.D.J.M. Kerkhof and J.A. Moulijen, J. Phys. Chem. 83:1612
    (1979).
20. C.H. Giles, T.H. MacEwan, S.N. Nakhwa and D. Smith,
    J. Chem. Soc. 3973 (1960).
21. C.H. Giles, A.P. D'Silva and I.A. Easton, J. Colloid Interface
    Sci. 47:766 (1974).
22. J.J. Kipling, "Adsorption from Solutions of Non Electrolytes,"
    Academic Press, New York (1965).
23. B.T. Bowman, Soil Sci. Soc. Am. Proc. 37:200 (1973).
24. L.R. Dole, Ph.D. Dissertation, VPI & SU, Blacksburg, VA
    (1972).

25. S.J. Gregg and K.S.W. Sing, "Adsorption, Surface Area and Porosity," Academic Press, New York (1967).

26. Pam Davar, M.S. Thesis, Virginia Polytechnic Institute and State Univeristy, Blacksburg, VA., 1980.

27. C.H. Giles, S.K. Jain and A.S.A. Hassan, Chem. & Ind. 629 (1955).

28. R.R. Colwell and T. Kaneko, Appld. Microbiol. 29: 269 (1975).

29. Y. Kang, J.A. Skiles and J.P. Wightman, J. Phys. Chem., 84: 1448 (1980).

30. J.H. Scofield, J. Electron Spectrosc. Related Phenom. 8: 129 (1976).

31. R.G. Nordberg et al., Arkiv Kemi 28: 257 (1968).

INTERACTION OF ZINC IONS WITH HYDROXYLAPATITE

D.N. Misra and R.L. Bowen

American Dental Association Health Foundation
Research Unit
Center for Materials Science
National Measurement Laboratory
National Bureau of Standards
Washington, D.C. 20234

ABSTRACT

The interaction between zinc ions in aqueous nitrate solutions and hydroxylapatite, the structural prototype for the principal inorganic constituent of tooth and bone, was studied. The nature of this interaction may be important in elucidating the role of zinc oxide commonly dissolved in the phosphoric acid solutions used to etch hard tooth tissues prior to the application of dental resins. This study has also a direct bearing on the suitability of zinc ions as candidate "mordants" for use with surface-active comonomers that promote adhesion between restorative resins and hard tooth tissues.

The reaction of zinc ions with hydroxylapatite diminishes with time even at dilute concentrations, and is not complete after several days at room temperature. X-ray analysis shows the formation of hopeite $[Zn_3(PO_4)_2 \cdot 4H_2O)]$ and the zinc analogue $[Zn_2(OH)PO_4]$ of libethenite. The initial phase of the interaction may be explained by two proposed rate processes occurring sequentially. One involves the exchange of the surface calcium ions with the zinc ions in solution and the other seems to depend inversely on the amount of adsorbed zinc ions. The rate constant of the second kinetic process is a linear function of the ionic strength of solution.

INTRODUCTION

Many interfacial[1-4] and adsorptive[5-9] properties of hydroxylapatite have been reported. The interaction of cupric and nickel-
ous ions with the apatite[10,11] has been studied. The interaction
of zinc oxide ions in aqueous solutions with hydroxylapatite, the
structural prototype for the principal inorganic constituent of
tooth and bone, is important for two reasons. Firstly, it may
elucidate the role of zinc oxide in dental etching solutions
where it is commonly dissolved in the phosphoric acid solutions
prior to the application of dental resins[12-14]. Secondly, this
interaction has a direct bearing on the suitability of zinc ions
as candidate "mordants"[10,11] for use with surface-active coupling
agents that promote adhesion between restorative resins.

The mechanism of reaction of zinc ions with hard tooth
tissues is different from that of other metal ions, e.g., Al, Cu,
Co or Fe. This is revealed when bovine teeth are allowed to
react with an excess of the aqueous nitrate solution of each ion.
Over a period of months, the teeth disintegrate in every solution
studied, except zinc nitrate, where they maintain their structural
integrity and strength. The reaction with metal ions other than
zinc results in complete replacement of calcium ions and formation
of metal phosphates.

The initial reaction of aqueous zinc nitrate with hydroxylapatite is important for estimating the rate and extent of zinc
ion exchange with the surface calcium ions. The "mordanting"
provides cationic binding sites for chelating monomers (coupling
agents) that in practice would be applied subsequently. This
exchange is complete within a few minutes if the concentration of
solution exceeds 0.2 mol/L. A rate-law explaining certain salient
features of this reaction was derived and is presented in a
following section.

A rate-law for the subsequent interaction between aqueous
zinc nitrate solution and hydroxylapatite, beyond the surface-
exchange but still pertaining to the initial phase of the reaction,
was also derived on the basis that the rate depends directly on
the concentration and inversely on the adsorbed amount of zinc
ions. The reaction products, as shown by X-ray analysis, are
hopeite [$Zn_3(PO_4)_2 \cdot 4H_2O$)] and the zinc analogue [$Zn_2(OH)PO_4$] of
libethenite [$Cu_2(OH)PO_4$].

EXPERIMENTAL

Materials*

The hydroxylapatite was Fisher certified tribasic calcium phosphate (C-127), with a chemical formula given as approximately $Ca_{10}(OH)_2(PO_4)_6$. It was repeatedly washed with boiling water before use; the physical and chemical details of its preparation and analysis have been described elsewhere.[10] It had a surface area (BET, N_2) of 41 m^2/g. The amount of physically adsorbed water on the apatite (1.57%) was determined by evacuating (at 100 N/m^2) the weighed samples at 105°C for several hours and then weighing after dry air was introduced into the vessel.

The zinc nitrate was "Baker analyzed" reagent-grade chemical, and Zincon (for the specific determination of zinc) was Fisher certified.

Methods

The apatite samples (1.000 g each) were shaken with 5.00 ± 0.05 mL of distilled water for two days at room temperature (23.0 ± 0.5°C) until a homogeneous slurry was obtained. A standardized metal nitrate solution (5.00 ± 0.05 mL) was added to each slurry sample and the mixture was shaken for a predetermined period and filtered through a medium-pore fritted disc. The concentration of zinc ions in the filtrate, after appropriate dilution, was determined spectrophotometrically by using the Zincon reagent.[15] The adsorbance was read at 620 nm against a reagent blank sample containing an amount of calcium nitrate similar to that of the filtrate. The amount of zinc ion uptake is equal to $V(C_i-C_f)/W$, where V(L) is the volume of solution in contact with W g of the apatite, and C_i (mol/L) and C_f (mol/L) are initial and final concentrations of solution. The rates of zinc ion uptake are determined for initial concentrations of 0.05, 0.10, 0.20 and 0.40 mol/L of zinc nitrate solution. The rates were also determined for two mixtures containing nitrates of zinc and calcium in respective molar ratios of: 0.15 to 0.10; and 0.20 to 0.20. Each rate point is an average of two or three separate experiments and is reproducible to within 10%.

The X-ray diffraction pattern for the product, obtained by

---

*Certain commercial materials are identified in this paper to specify the experimental procedure. In no instance does such identification imply recommendation or endorsement by the National Bureau of Standards or the ADA Health Foundation or that the material identified is necessarily the best available for the purpose.

reacting hydroxylapatite with an excess of concentrated solution
of zinc nitrate at boiling temperature for several weeks, showed
peaks at 2 θ values corresponding to those of all the strong
hopeite [$Zn_3(PO_4)_2 \cdot H_2O$] and libethenite [$Cu_2(OH)PO_4$] lines. No
data for $Zn_2(OH)PO_4$ are available. No other strong peaks were
observed, and for hopeite the intensities were similar to the
standard hopeite patterns. Concentrations of less than 5% of
contaminants would not be readily detectable, but strong lines of
other compounds should show up. Further comparative data are
given in Table 1 demonstrating that the product is a mixture of
hopeite and zinc analogue [$Zn_2(OH)PO_4$] of libethenite [$Cu_2(OH)PO_4$].
The strong lines of adamite [$Zn_2(OH)AsO_4$] match poorly with those
of the product.

Results

The uptake of zinc ions by hydroxylapatite from four different
aqueous nitrate solutions is shown in Figures 1 and 2 for two
time-periods. Figure 3 shows the uptake for two zinc nitrate
solutions and for two zinc nitrate mixtures with calcium nitrate
during the early period of reaction. Whereas explanations for
the fast surface ion-exchange process (up to 1-5 h) (Figure 3) and

Table 1.   Comparison of Some Strongest X-Ray Powder
            Diffraction Lines for Hopeite, Libethenite
            and the Reaction Product of Hydroxylapatite
            with Aqueous Zinc Nitrate Solution

Hopeite[a]		Libethenite[b]		Reaction Product[c]	
d	I	d	I	d	I
9.16	65	–	–	9.14	68
4.56	40	–	–	4.58	84
2.85	100	–	–	2.86	100
–	–	5.85	90	5.82	10
–	–	4.81	100	4.82	10
–	–	2.40	60	2.41	4

[a] $Zn_3(PO_4)_2 \cdot 4H_2O$

[b] $Cu_2(OH)PO_4$

[c] In an incompletely reacted mixture at room temperature,
     the hopeite crystals are visible in the product as thin
     rectangular birefringent blades with refractive indices
     of 1.588 and 1.596.

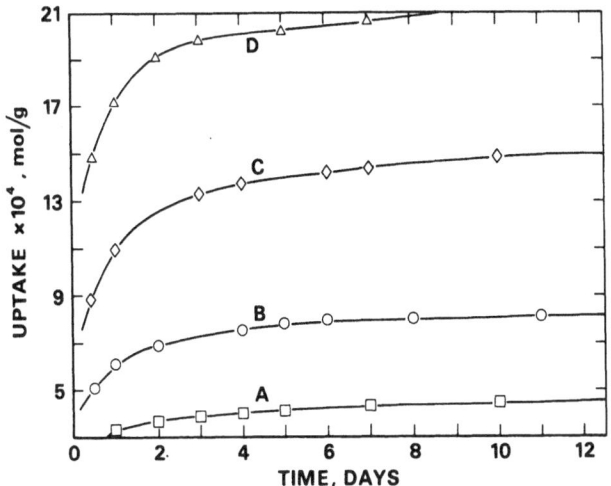

FIGURE 1   Uptake of zinc ions by hydroxylapatite at 23°C for a
period of about two weeks from aqueous nitrate solutions
having initial concentrations (mol/L) of: (A) 0.05,
(B) 0.10, (C) 0.20, and (D) 0.40.

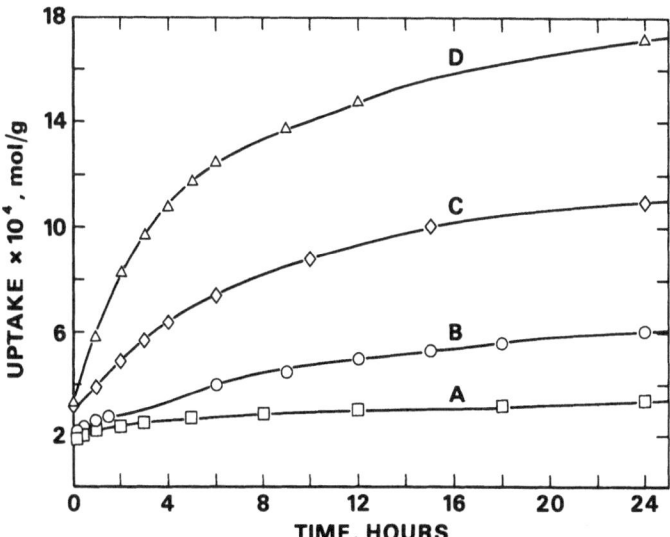

FIGURE 2   Uptake of zinc ions by hydroxylapatite at 23°C for a
period of 24 hours from aqueous nitrate solutions having
initial concentrations (mol/L) of: (A) 0.05, (B) 0.10,
(C) 0.20, and (D) 0.40.

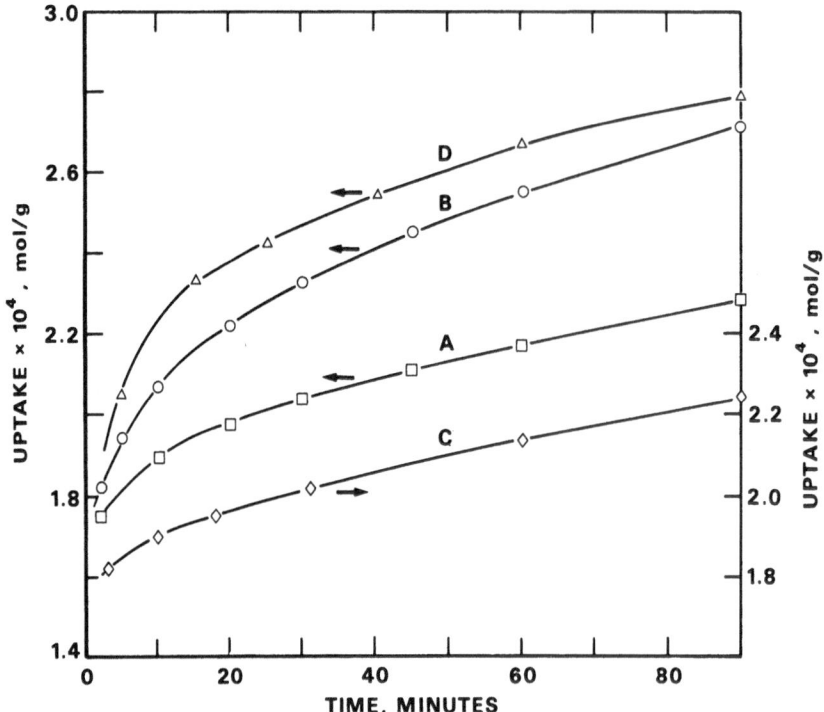

FIGURE 3   Uptake of zinc ions by hydroxylapatite at 23°C for a
            period of 90 minutes from aqueous nitrate solutions having
            initial concentrations (mol/L) of: (A) 0.05, (B) 0.10,
            (C) 0.15, and $[Ca(NO_3)_2]$ = 0.10, and (D) 0.20, and
            $[Ca(NO_3)_2]$ = 0.20.

the subsequent uptake of zinc ions (up to 18 h) (Figure 2) have
been developed in the next section, no such theoretical attempt
has been made for the reaction continuing for longer periods
(Figure 1).

The uptake of zinc ions by hydroxylapatite is related to the
release of calcium ions into the solutions (Table 2). There is a
one to one correspondence within the limits of experimental
errors (within 10%). The average pH during the time interval was
about 3.8 ± 0.1 and it may have a bearing on the slightly higher
calcium ion concentration values.

DISCUSSION

The maximum amount of exchangeable calcium ions on hydroxylapatite, as determined by ion-exchange adsorption and initial
rates of uptake of Ni and Cu, is 3.00 x 10^{-4} mol/g.[10,11] The
rates of zinc ion uptake, like those of Ni and Cu,[11] during the
early period of reaction (Figure 3), may be represented by the
equation:

Table 2.   Uptake of Zinc Ions vs. Release of Calcium
            Ions[a] into Solution during the Reaction of
            Hydroxylapatite (1 g) with Zinc Nitrate
            (10 mL, 0.20 mol/L)

Time day	(Zn^{2+} uptake) x 10^4 mol/L	(Ca^{2+} release) x 10^4 mol/L
1	11.0	11.5
2	12.6	13.1
5	14.0	14.5
10	14.9	15.2

[a]   Calcium ion concentrations were determined in the
       presence of zinc ions by atomic absorption spec-
       troscopy.

$$-\frac{dC}{dt} = kC(q_o-q) \qquad\qquad (1.a)$$

but, since $CV = C_oV - q$,

$$\frac{dq}{dt} = k\,(C_oV - q)\,(q_o-q) \qquad\qquad (1.b)$$

where C is the zinc ion concentration at any time t; C_O is the
initial concentration of zinc ions in the solution; q is number
of moles of zinc ion taken up per gram of apatite at any time t;
q_o is the maximum value q would have if all the exchange sites
were occupied; V is the volume of solution in contact with 1 g
of the apatite; and k is the specific ion exchange rate constant.
Integration of Eq. (1.b) yields:

$$\ln\left(\frac{C_oV-q}{q_o-q}\right) = k\left(C_oV-q_o\right)t + \ln\frac{C_oV}{q_o} \; ; \qquad\qquad (2)$$

where the second term on the right-hand side is the integration
constant and is evaluated from the initial condition: i.e.,
q = 0 when t = 0.

    If the expression on the left-hand side of Eq. (2) is plotted
against t (Figure 4), the slope gives the rate constant, k =
slope/(CV-q_o), and the intercept, $\ln(C_oV/q_o)$, should be related
to the experimental constants. The presence of Ca ions in

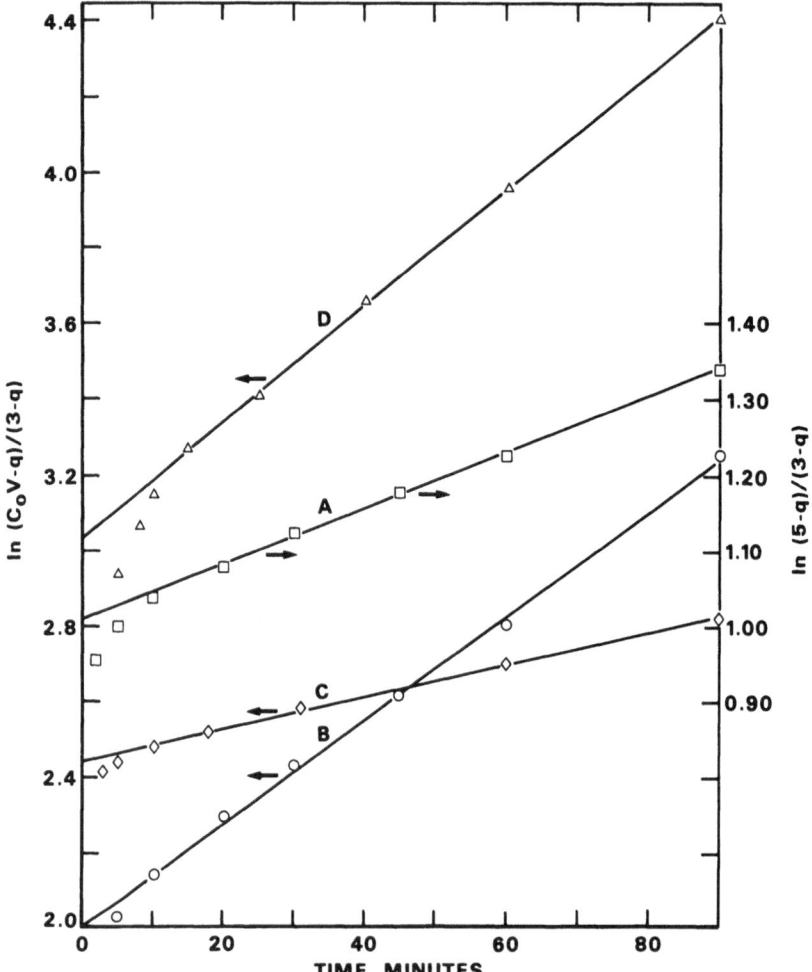

FIGURE 4   Rate-law plots for surface-exchange of zinc ions at 23°C
           on hydroxylapatite from aqueous zinc nitrate solutions
           having initial concentrations (mol/L) of: (A) 0.05,
           (B) 0.10, (C) 0.15 and $[Ca(NO_3)_2] = 0.10$, and (D) 0.20
           and $[Ca(NO_3)_2] = 0.20$.

solution would influence the specific rate constant for zinc ion
uptake and may be accounted for by introducing a modifying factor
in Eq. (1), which now becomes:

$$-\frac{dC}{dt} = kC\left[(q_o-q) - AC_{Ca}(q_o-q)\right],$$
(3)

where A(L/mol) is a constant and C_{Ca} is the concentration of ions
at time t.   In the absence of added Ca ions, C_{Ca} is very small
and is equal to the Ca ions released into the solution by ion-
exchange process at the surface.   The whole modifying factor may,

therefore, be neglected. In the presence of added Ca ions having an initial concentration of C_1, $C_{Ca} \simeq C_1$. Integration of Eq. (3) yields:

$$\ln \left(\frac{C_o V - q}{q_o - q} \right) = k(1 - AC_1)(C_o V - q_o)t + \ln \frac{C_o V}{q_o} .$$
(4)

Now, when $\ln((C_o V - q)/(q_o - q))$ is plotted against t, the rate constant (k) = slope/$(1 - AC_1)(C_o V - q_o)$ but the slope remains the same as that of Eq. (2).

In the very early phase of the reaction (till about 10 min), the hydrodynamic factors, e.g., the rate of mixing etc., would probably control the rate of reaction. Thereafter, the rate-law plots (Figure 5) are fairly linear until about 80% of the ion-exchange process at the surface is complete. The slopes, derived rate constants, and the intercepts are presented in Table 3. The two values of the rate constant agree satisfactorily, but the experimental and calculated values of the intercepts do not. Both the experimental and calculated values for the intercept increase with the concentration of zinc ions. The ratio of the two values (experimental/calculated) is inversely related to the zinc ion concentration in the solution.

The disagreement between the experimental and the calculated values of the intercepts show that the representation of the initial zinc ion-exchange rate process by Eq. (1), like that of the case of cupric ions[11] is decidedly incomplete. However, to keep the rate-law simple and manageable, without introducing other indeterminable rate constants, various factors, e.g., the desorption kinetics of zinc and calcium ions, the influence of pH and the ionic strength of the solutions, etc., were not taken into account. Also, deviation from Eq. (2) or (4) increases with time, during which a slow chemical reaction takes place. Zinc ions are removed from solution with the introduction of an equiva-lent amount of calcium ions and the hopeite becomes visible as birefringent crystals under a polarizing microscope. The kinetics of the initial ion exchange uptake, however, is approximated by Eq. (1) since the individual plots are quite linear after the initial mixing time (~ 10 min) and the two values of the rate constant are in fair agreement.

If the rate constant (k) for zinc ions is assumed to be the same (19.51 g/mol.min), even if the calcium ions are present initially, the respective values of A(L/mol) for Ca ions can be calculated as 8.18 and 2.17. These values of A depend inversely on hydrogen-ion concentration of the solution (i.e., $A[H^+] = 1.27$, and 1.19 respectively). The surface exchange of zinc ions is

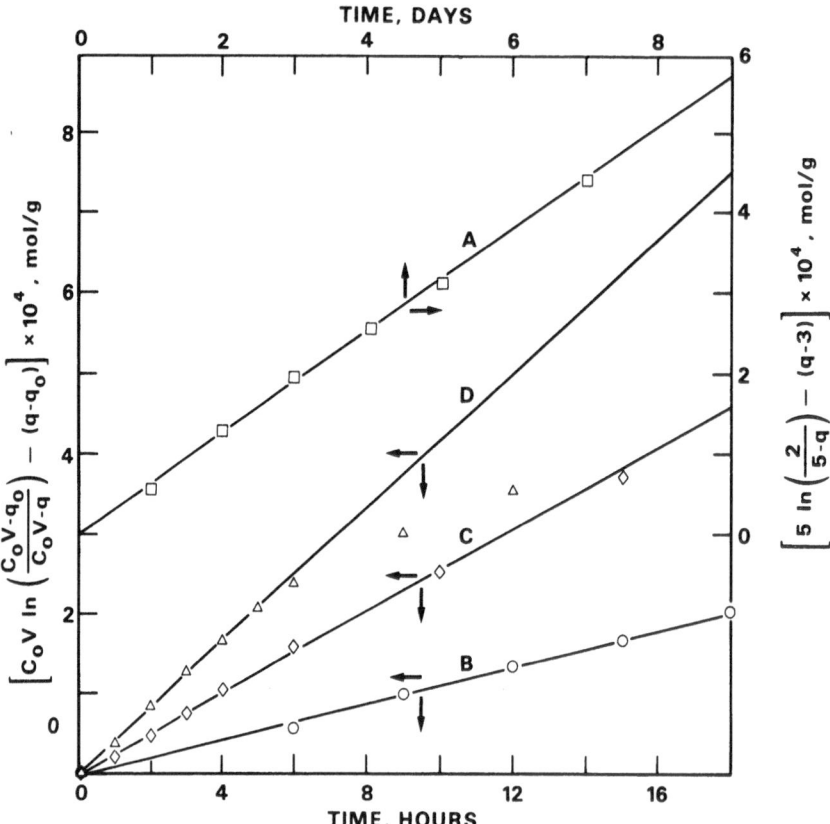

FIGURE 5   Rate-law plots for uptake of zinc ions (after initial
            surface-exchange) at 23°C on hydroxylapatite from aqueous
            nitrate solutons having initial concentrations (mol/L)
            of: (A) 0.05, (B) 0.10, (C) 0.20, and (D) 0.40.

complete within a few minutes, if the initial concentration of
zinc exceeds 0.2 mol/L.

An optical examination of the apatite samples shows the
presence of birefringent blades of hopeite, which develop, depending on the starting concentration of zinc nitrate solution, in a
period of 4 to 18 hours. The optical examination, also, shows
that the number of fresh reaction sites does not increase as the
reaction proceeds. This observation supports a hypothesis that
the surface of hydroxylapatite, in the initial stages of the
reaction, is coated with zinc analogue of libethenite, whereas
hopeite forms another phase. The kinetics of this phase of the
reaction may be represented as:

$$-\frac{dC}{dt} = k_2 \frac{C}{q}, \quad \text{or} \quad \frac{dq}{dt} = k_2 \frac{(C_o V-q)}{q} \qquad (5)$$

Table 3.   Initial Ion-Exchange Rates of Zinc on Hydroxylapatite[a]
            at 23°C

Initial Concentr. mol/L [Zn^{2+}] [Ca^{2+}]		pH[b] Soln. Slurry		Slope x $10^{3,c}$	Rate constant[d] (k) g/mol.min	Intercept[c] Exptl. Calcd.	
0.05	-	5.37	4.10	3.73	18.7	1.01	0.51
0.10	-	5.18	3.97	13.66	19.5	2.01	1.20
0.15	0.10	5.78	3.81	4.26	-	2.44	1.61
0.20	0.20	5.79	3.35	15.49	-	3.02	1.90

[a]   The maximum amount of ion-exchange uptake is 3.00×10^{-4} mol/g.

[b]   The pH was determined 10 min after 1 g of apatite was added to
       10 mL of a zinc nitrate solution. In each case, the rise in
       the pH in slurry is less than 0.02 unit after 24 h. The pH of
       the slurry of apatite (1 g) in distilled water (10 mL) was 6.83.

[c]   Experimental values are obtained from Figure 4. The correlation
       coefficient in each case, for the values determined 10 min or
       after, is greater than 0.99.

[d] $k = $ slope/(C_oV-q_o). If k is assumed to be 19.51 g/mol·min for
       the last two cases (when Ca ions were present initially), the
       respective values of A (L/mol) can be calculated as 8.18 and
       2.67. These values of A depend inversely on [H$^+$] of the solu-
       tion (i.e., A [H$^+$] $\cong$ const.).

where k_2 is the specific reaction rate constant and other symbols
have the usual meaning. Integration of Eq. (5) gives:

$$k_2 t = C_o V \ln \frac{C_o V - q_o}{C_o V - q} - (q - q_o) . \qquad (6)$$

Eq. (6) includes the integration constant which is evaluated from
the initial condition: i.e., $q = q_o$ when $t = 0$.

    If the expression on the right-hand side is plotted against
t (Figure 5), the linear curve should pass through the origin and
the slope is equivalent to the specific rate constant k_2. The
plots are fairly linear (Figure 5) if the uptake of zinc ions is
limited to about 10×10^{-4} mol/g or less. The amount of zinc

Table 4.   Early Rate of Uptake of Zinc Ions[a] on Hydroxyl-
            apatite at 23°C

Initial $[Zn^{2+}]$ mol/L	Slurry pH[b]	Ionic Strength, M mol/L	Slope[c] or k_2 x 10^5 mol/g·h
0.05	4.1	0.15	0.26
0.10	4.0	0.30	1.11
0.20	3.7	0.60	2.56
0.40	3.5	1.20	4.17

[a]   After the initial ion-exchange process is complete
       at the surface with an uptake of 3.00 x 10^{-4} mol/g
       of zinc.

[b]   See footnote (b) of Table 3.

[c]   The correlation coefficient in each case, for the
       values determined up to 10 x 10^{-4} mol/g (or less)
       of zinc ion uptake, is greater than 0.99.

ions in the first unit cells (area = U = 66.5 $\overset{\circ}{A}^2$, containing 10
Ca ions) of the hydroxylapatite surface (area = S = 41 m^2/g) may
be easily calculated and is equal to 10 S/NU(= 10.24 x 10^{-4} mol/g),
where N is Avogadro's number. It appears, therefore, that the
rate-law is valid only if the reaction remains limited to the
first surface unit cells of hydroxylapatite. Possibly, the zinc
analogue of libethenite begins to form a separate phase at this
stage, and the assumptions implied in the rate-law Eq. (5) are
invalidated. A study is now underway to determine whether or not
the development of the libethenite (zinc analogue) coating plays
any role in the preservation of the structural integrity of
bovine teeth in concentrated zinc nitrate solutions.

    The specific rate constant (k_2) seems to depend on the ionic
strength of the solution or the pH of the slurry (Table 4).
Fairly linear plots are obtained in both cases. The plot k_2 vs.
the ionic strength of the solution results in a straight line
passing through the origin (correlation coefficient ~ 0.99)
(Figure 6). Further work is required to develop a theoretical
understanding of these correlations.

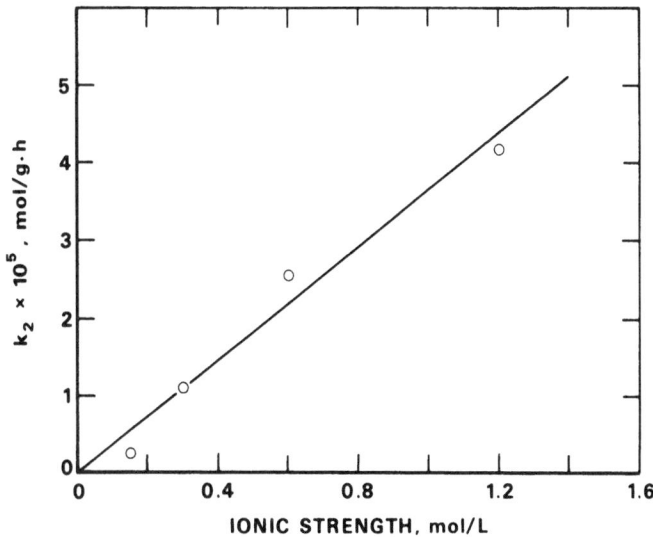

FIGURE 6   Rate-constant (k_2) vs. ionic strength of solution at 23°C.

ACKNOWLEDGEMENTS

     This investigation was supported, in part, by Research Grant
7 R01 DE05129-03 to the American Dental Association Health Foundation from the National Institutes of Health - National Institute
of Dental Research, and is part of the dental research program
conducted by the National Bureau of Standards in cooperation with
the American Dental Association Health Foundation.

     The authors thank Mr. C.P. Mabie for optical examination of
the product.

REFERENCES

1. P. Somasundaran, J. Colloid Interface Sci. 27:659 (1968).
2. L.C. Bell, A.M. Posner and J.P. Quirk, J. Colloid Interface
    Sci. 42:250 (1973).
3. A.N. Smith, A.M. Posner and J.P. Quirk, J. Colloid Interface
    Sci. 48:442 (1974).
4. S. Chander and D.W. Fuerstenau, J. Colloid Interface Sci.
    70:506 (1979).
5. M.A. Spinelli, F. Brudevold and E. Moreno, Archs. Oral Biol.
    16:187 (1971).
6. D.N. Misra and R.L. Bowen, J. Colloid Interface Sci. 61:14
    (1977).
7. D.N. Misra and R.L. Bowen, J. Phys. Chem. 80:842 (1977).

8. M. Kresak, E.C. Moreno, R.T. Zahradnik and D.I. Hay, <u>J. Colloid Interface Sci.</u> 59 (1977).

9. V. Hlady and H. Füredi-Milhofer, <u>J. Colloid Interface Sci.</u> 69:460 (1979).

10. D.N. Misra, R.L. Bowen and B.M. Wallace, <u>J. Colloid Interface Sci.</u> 51:36 (1975).

11. D.N. Misra and R.L. Bowen, <u>J. Biomed. Matls. Res.</u> 12:505 (1978).

12. Z. Sheykholeslam and M.G. Buonocore, <u>J. Dent. Res.</u> 51:1572 (1972).

13. H.W. Snyder, C.E. Wilson, G.V. Newman and J. Semen, <u>J. Appl. Poly. Sci.</u> 11:1509 (1967).

14. D. Macko, M. Rutberg and K. Langeland, <u>J. Dent. Res. (IADR Abs.)</u> 55:B228 (1976).

15. F.D. Snell, "Photometric and Fluorimetric Methods of Analysis (Metals)", Part 2, J. Wiley & Sons, New York (1978). pp. 1063-64).

SORPTION OF ACTINIDES FROM AQUEOUS SOLUTIONS UNDER

ENVIRONMENTAL CONDITIONS

G.W. Beall[1] and B. Allard[2]
Oak Ridge National Laboratory
Oak Ridge, TN 37830

ABSTRACT

The sorption of americium, neptunium, and plutonium has been studied on a suite of thirty pure minerals and one granite as a function of solution pH, salinity, and redox potential. Three major factors controlling sorption have been identified. These three major factors are pH (for hydrolyzable ions), chemi-sorption reactions at the surface of minerals that contain strong complexing ions, and redox reactions with ferrous iron contained in various minerals. It appears that simple cation exchange does not play a significant role in the pH range of most interest in groundwaters. The implications of these findings for nuclear waste disposal are discussed.

INTRODUCTION

After initial dominance of fission products, the actinides and their daughter products (americium, plutonium, neptunium, thorium, radium) would dominate the biological hazards from high-level reprocessing wastes as well as from unreprocessed spent uranium fuel from about 300 years after the discharge from the reactor up to millions of years. Thus, an understanding of the chemical behavior of the actinides in nature is of major impor-tance, e.g. for the design and safety analysis of an underground repository for high-level nuclear wastes.

---

1 Current address: Radian Corporation, 8500 Shoal Creek Blvd., Austin, Texas 78766
2 Current address: Department of Nuclear Chemistry, Chalmers University of Technology, S-41296 Göteborg, Sweden.

Most literature data on actinide retention in rock/ground-water systems under environmental conditions indicate high sorption for species in the tri- and tetravalent states (distribution coefficients, mol/kg rock per mol/m^3 water, above 1 m^3/kg) and generally low values for species in the penta- and hexavalent states (distribution coefficients below 0.1 m^3/kg). However, measurements in similar systems have often given contradictory results.[1]

The qualitative and quantitative effects on the retention of actinides by the variation of chemical conditions such as the reduction-oxidation potential, pH, the composition of the ground-water, the mineralogical composition of the ground, etc., are far from well understood. For this study, the sorption of americium, neptunium and plutonium has been measured under simulated environmental conditions to get qualitative information on the relative importance of pertinent chemical parameters on the actinide sorption characteristics in igneous bedrock.

EXPERIMENTAL

The sorption of neptunium,[2] plutonium and americium[2-4] was measured on selected minerals and rocks, using a batch technique.[2] Pure minerals were crushed in a mortar and sieved. The purity of these minerals was primarily determined by X-ray diffraction. Only the size fraction 0.044-0.063 mm was used in the sorption experiments. The cation and anion exchange capacities (with respect to Na$^+$ and Br$^-$) were measured by a batch technique,[5] and the specific surface areas were determined by the ethylene glycol method.[6] The crushed and sieved solid was prewashed and pre-equilibrated with an artifical groundwater solution (see Table 1). Active acidic spike solutions were added, giving a total initial radionuclide concentration of 2.0 x 10^{-11} M for neptunium (^{235}Np), 1.8 x 10^{-11} M for plutonium (^{237}Pu), and 2.1 x 10^{-9} M for americium (^{241}Am). The uptake of the radionuclide on the solid sorbent (solid-liquid ratio of 7-12 g/L) was measured after a contact time of 5 d at ambient temperature. Experimental details concerning spiking, sampling and counting procedures are given elsewhere.[2]

GENERAL DISCUSSION

The Bedrock/Groundwater System

Igneous rocks like granite, basalt etc., are largely composed of a small number of primary, rock-forming minerals (see Figure 1) and some additional accessory minerals, as well as weathering and decomposition products of these constituents. The complicated

Table 1.   Composition of artificial groundwater.

Species	Concentration in Nature, mg/L [1]	Standard Water, ORNL, mg/L
HCO_3^-	60 - 400	123
SiO_2 (total)	5 - 60	12
SO_4^{2-}	3 - 40	9.6
Cl^-	5 - 50	70
Ca^{2+}	10 - 60	18
Mg^{2+}	2 - 25	4.3
K^+	1 - 10	3.9
Na^+	10 - 100	65
F^-	0.01 - 5	3.8 [2]
HPO_4^{2-}	0.01 - 0.5	0.89 [2]
Fe^{2+}	0.5 - 20	2 [2]
pH	7.2 - 8.5	8.2

[1] Probable concentration range for undisturbed deep groundwater in contact with igneous rocks[7,9,31].

[2] Not added in standard batch.

interrelations and exchanges between species in solution and the minerals of the igneous rock are illustrated in Figure 2.

Complex Forming Anions.   Carbonate in groundwaters [CO_3^{2-}-HCO_3^--$H_2(CO_3)$], which largely originates from carbon dioxide in the atmosphere and biosphere, acts as a pH-buffer, giving a pH of 8 to 9 in an undisturbed water.   The total carbonate concentrations are usually high (in the 1-7 mM range) and always significant even in totally isolated systems.   Among other anions in groundwaters, the contents of phosphate (PO_4^{3-}-HPO_4^{2-}-$H_2PO_4^-$) and fluoride (F^-) may both be related to the calcium concentration which is limited by the solubility product of calcium carbonate.

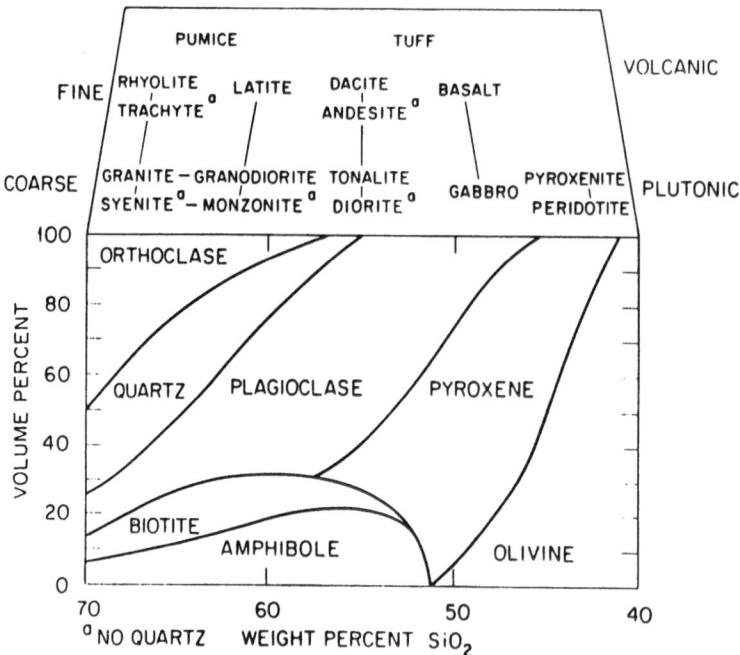

FIGURE 1: The major minerals of igneous rocks.[2]

Although the phosphate and fluoride concentrations are low, the presence of these strongly complexing ions in the water should not be overlooked. Independent of the carbonate–pH system, but related to the contact with the atmosphere and biosphere, is the sulfate content of the groundwater. In Table 1 some representative anion concentration levels for granite groundwaters are summarized.

The concentrations of complexing organics like humic and fulvic acids are low in deep groundwaters[7] but might be substantial in surface waters.

Reduction-Oxidation Properties. The reduction-oxidation potentials in groundwaters are largely determined by the presence of air or, in deep waters without immediate contact with the atmosphere, are determined and buffered by the presence of ferrous (Fe(II)) minerals like pyrite (FeS), magnetite (Fe_3O_4), and others. The redox potential (E) of an aerated water would roughly be given by

$$E = 0.8 - 0.06 \text{ pH (V)} , \tag{1}$$

giving a value of about 0.3 V at pH 8.[8] In the presence of ferrous minerals at expected concentration levels, the redox potentials in granite groundwater would be estimated from

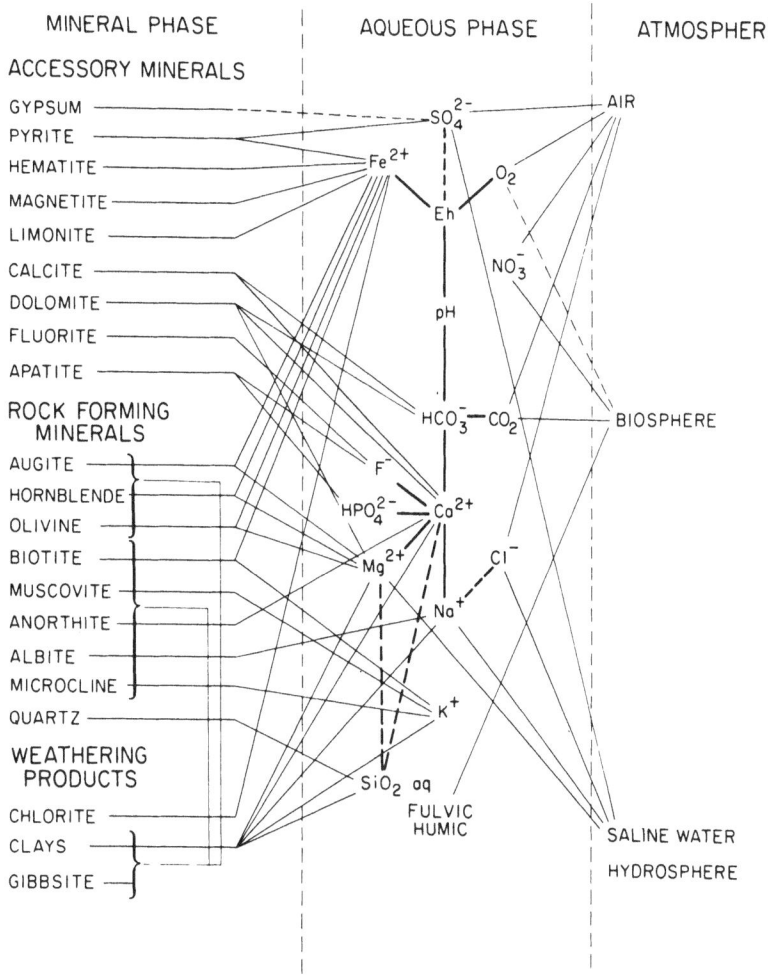

FIGURE 2: The bedrock/groundwater system of igneous rocks.[3]

$$E = 0.2 \pm 0.1 - 0.06 \, pH \, (V) , \qquad (2)$$

giving values in the range of -0.3 to -0.4 V at pH 8.[9,10]
Negative redox potentials down to this level have been observed
in situ.[11]

Actinide Chemistry in Natural Waters

   Complex Formation.   The actinides form strong complexes of
the hard-hard type[12] with many oxygen-containing anions in natural
waters (hydroxide, carbonate, phosphate, sulfate) as well as with
fluoride.[13]   The strength of the complex varies considerably

Table 2. Expected actinide complex formation constants (25°C, ionic strength 0) for the reaction $An + L \rightarrow AnL$, K_1, and solubility products K_S (for U, Np, Pu and Am)[13,14,18,32,33]

Ligand L	Am(III)	Am(IV)	Am(V)	Am(VI)
		$\log K_1$		
OH^-	6.6-8	13.2-13.7	4.0-5.0	8.1-9.1
CO_3^{2-}	≥ 8			12
NO_3^-	1.0-1.2	1.6-1.8	0.1	0.8-1.0
		10.8-12.9	2.9	3.4
SO_4^{2-}	3.5-3.8	5.5-5.8	2	2.9-3.2
F^-	3.6-4.3	8.5-9.2	3.7	4.5-4.8
Cl^-	1.0-1.2	0.9-1.3	< 0	0.2-0.4
Humic acid[1]	6.8	12-13		5.8

[1] Binding constants, 0.1 M $NaClO_2$, for Am(III), Th(IV) and U(VI) and pH 4.5-5[34].

Ligand L	Am(III)	Am(IV)	Am(V)	Am(VI)
		$- \log K_S$		
OH^-	24-25	53-55	9	22-23
CO_3^{2-}	30-33			12
PO_4^{3-}	23	> 57	1	47-50
F^-	15-16	19-26		

[1] Low solubility, K_S values not available.

between the different valence states, usually with a decrease in strength in the order of An(IV) > An(VI) $\simeq$ An(III) > An(V). In Table 2 some measured and estimated complex constants are given, which indicate the relative importance of the anions of natural waters for actinide complexation and speciation.

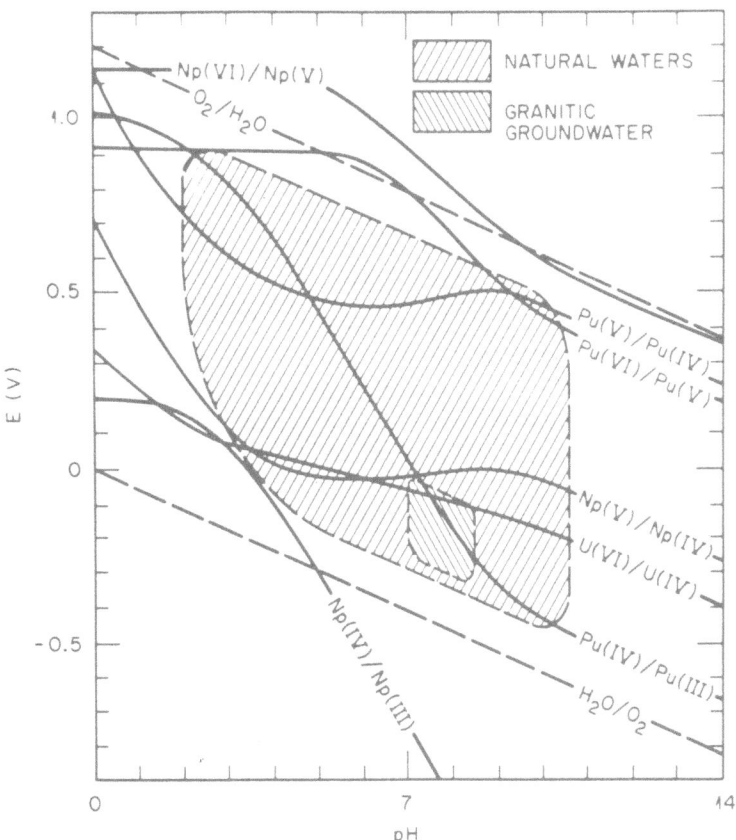

FIGURE 3: Potential-pH diagram for U, Np, Pu and Am at low concentrations and considering expected hydrolysis reactions.[14]

Valence States. Very few measurements of equilibria between
species of different valence states have actually been performed
other than in 1 M acidic solutions. From measured potentials at
pH 0, and considering hydrolysis, using measured or estimated
hydrolysis constants, potential-pH-equilibrium curves have been
generated for the reduction-oxidation sensitive actinides uranium,
neptunium, plutonium, and americium[14] (see Figure 3). Some
potential-pH-concentration data available for uranium bearing
rock formations support the proposed U(VI)/U(IV) curve presented
in Figure 3.[15,16]

Actinide Species in Groundwater. Considering the concentrations of inorganic complexing anions and pH, the actinides would
be expected to exist as the following species (at low concentrations) in groundwaters.

Americium. Americium would be solely trivalent in natural
waters and would predominantly exist as hydrolyzed species like
Am(OH)$^{3-x}$ and possibly also as carbonate complexes. At the
highest expected environmental fluoride concentrations, a substantial part of the americium in solution would exist as AmF^{2+}.

Plutonium. Dominating valence state in a closed groundwater
system would be Pu(IV) although any one of the other valence
states (III, V, or VI) could dominate under suitable conditions
(see Figure 3). For tetravalent species, hydrolysis would dominate
entirely over any other complex formation, giving possible
aqueous species like Pu(OH)$_x^{4-x}$, where x $\geq$ 2-4, depending on the
pH. A dehydration of species with many hydroxyl groups seems
feasible, leading to the eventual formation of hydrous oxides or
polymerization at sufficiently high concentrations.[17]

Neptunium. The tetravalent states would dominate in deep
groundwaters, and hydrolyzed products would be formed (c.f.
plutonium). Under oxic conditions, anionic carbonate complexes
of hexavalent uranium like UO$_2$(CO$_3$)$_3^{4-}$ would dominate in natural
waters.

EXPERIMENTAL RESULTS

Sorption of Actinides on Geologic Media

Influence of Hydrolysis. Among the chemical parameters that
were varied during the sorption experiments (concentrations of
anions in the aqueous phase, pH, composition of the solid phase),
the pH of the solution had the strongest influence on the sorption
under otherwise similar conditions. In Figure 4, the sorption of
Am(III), Pu(IV), and Np(V) on some low capacity minerals is given
as a function of pH, as well as the calculated relative concentrations of soluble species, considering hydrolysis reactions.[14,18]
The total radionuclide concentrations in these experiments are
low enough that precipitation of hydroxides in the bulk solution
would not be expected.[14,18,19]

A qualitative correlation between the observed sorption
isotherms and the calculated hydrolysis curves is obtained. A
substantial sorption can be noticed starting in the pH range
where significant hydrolysis would be expected (around pH 4 for
Am(III) and pH 7 to 8 for Np(V)). A slight decrease of the
sorption is observed for Am(III) at pH above 7 to 8, where a
significant fraction of the americium would exist as anionic
species. For Pu(IV) the sorption is high in the whole pH range
studied, even in the pH range where neutral or anionic species
would be expected to dominate. The change of cation and anion

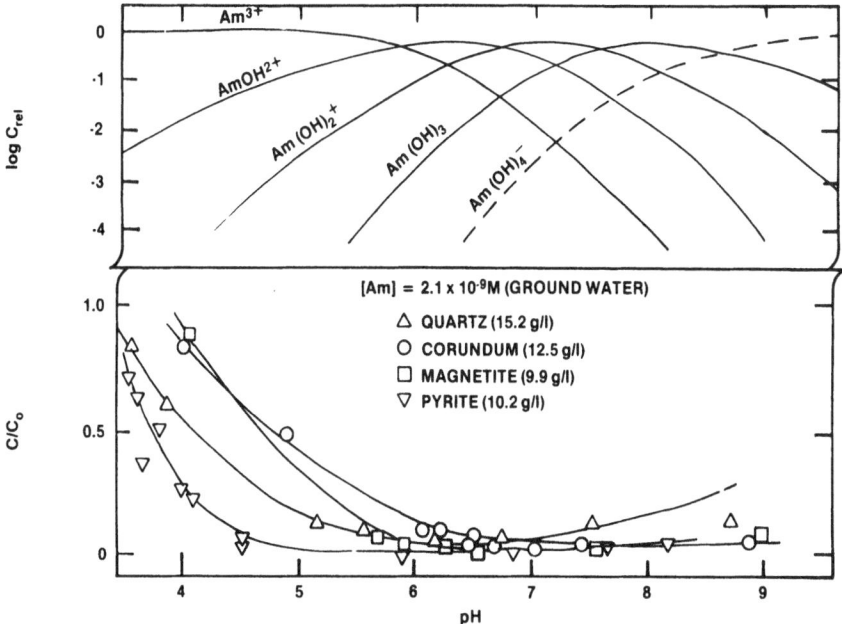

FIGURE 4a: Sorption of Am(III) on some low capacity minerals
              (C/Co) and calculated relative concentrations of
              hydrolyzed species as a function of ph.

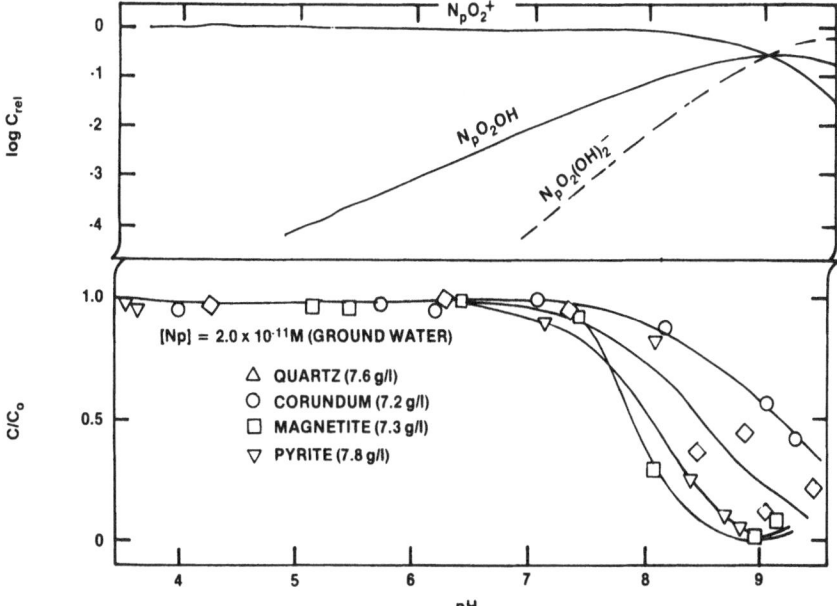

FIGURE 4b: Sorption of Np(V) on some low capacity minerals
              (C/Co) and calculated relative concentrations of
              hydrolyzed species as a function of ph.

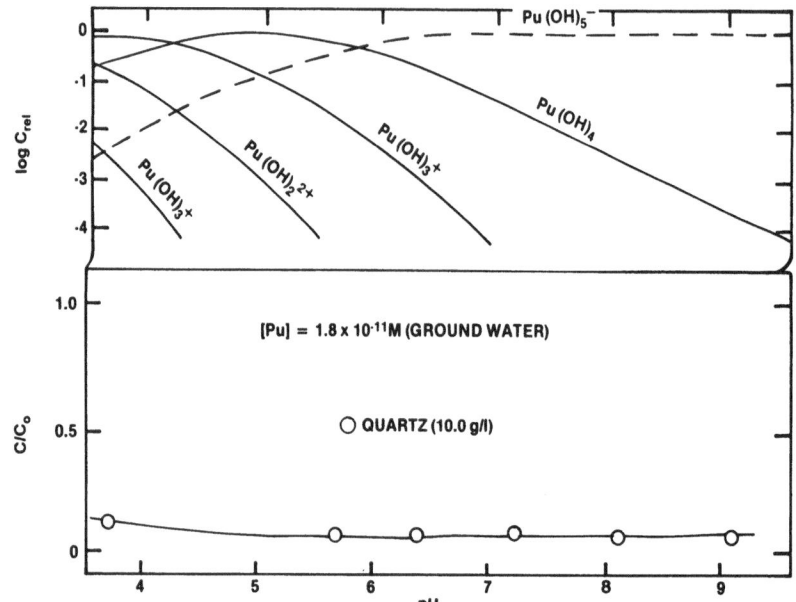

FIGURE 4c: Sorption of Pu(IV) on some low capacity minerals
                  (C/Co) and calculated relative concentrations of
                  hydrolyzed species as a function of ph.

exchange capacities with pH of the minerals of Figure 4 cannot
account for either the observed pH-dependence or the magnitude of
the sorption.[20]   However, the mineral composition somewhat affects
the pH where the sorption increase starts (within a range of 1 pH
unit around pH 4 for Am and around 8 for Np).

     The observed pH dependence is in agreement with the majority
of previous observations of the sorption of hydrolyzable radionuclides on inert solids at trace concentrations.[21-23] Generally,
a sorption increase with hydrolysis would be expected, giving a
maximum sorption in the pH range where neutral hydroxy compounds
(ion pairs) are formed and a decrease of the sorbed fraction at
higher pH, where a formation of anionic species would be feasible.
This general behavior has been studied in some detail for a
number of hydrolyzable transition metals,[24-26] and a sorption
model based on hydrolysis and complex formation has been proposed.

     Chemisorption by Complexing Anions.   From the formation
constants in Table 2 it seems evident that the presence of
strongly complexing anions like CO_3^{2-}, PO_4^{3-}, and F^-, either in the
aqueous phase or in the solid phase, would strongly affect the
sorption behavior of the actinides. Depending on the concentrations of these anions, various species would exist in solution.

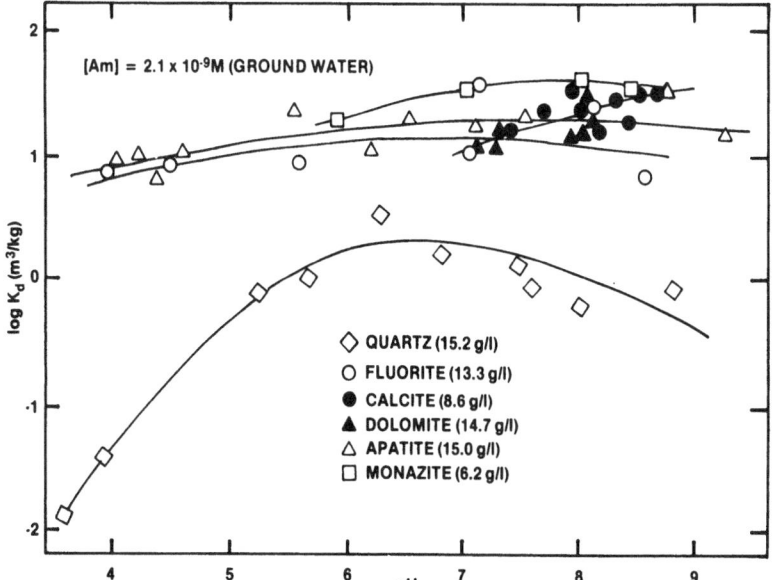

FIGURE 5a: Distribution coefficients (K_d) for Am(III) on some
                  phosphate, carbonate, and fluoride in contrast to
                  quartz-containing minerals.

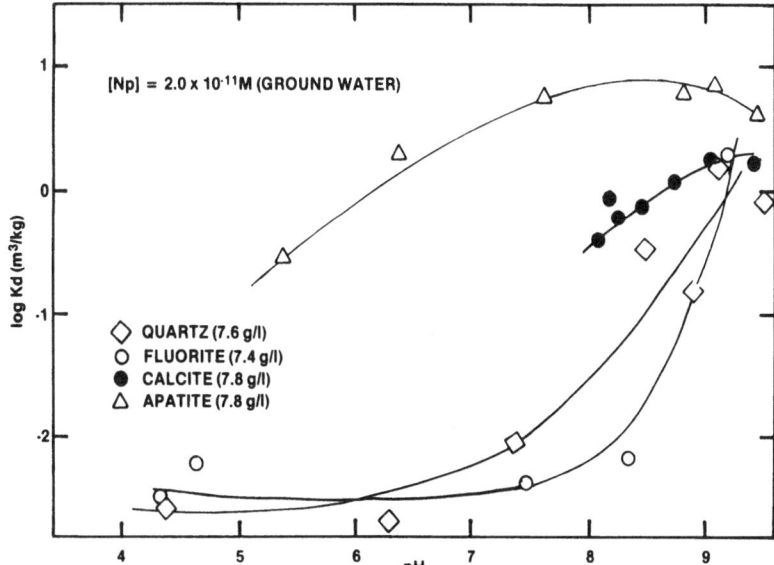

FIGURE 5b: Distribution coefficients (K_d) for Np(V) on some
                  phosphate, carbonate, and fluoride in contrast to
                  quartz-containing minerals.

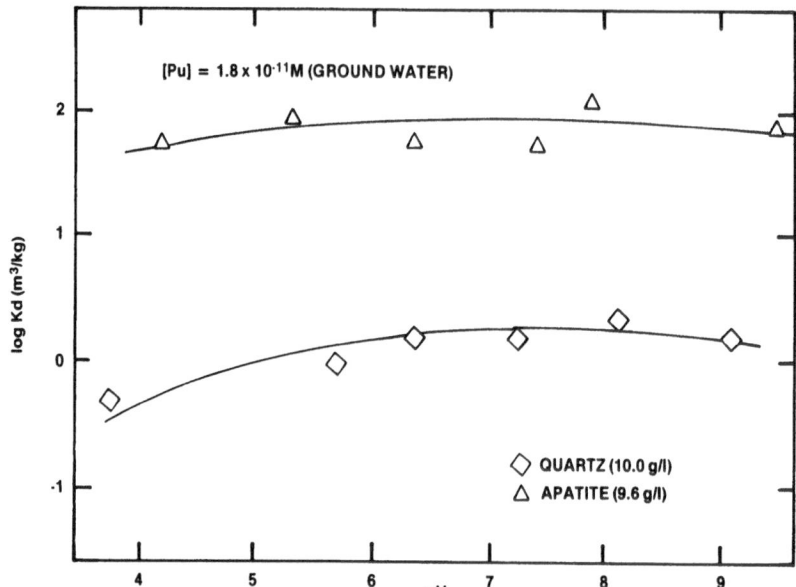

FIGURE 5c: Distribution coefficients (K_d) for Pu(IV) on some
                phosphate, carbonate, and fluoride in contrast to
                quartz-containing minerals.

For minerals containing CO_3^{2-} (calcite, dolomite), PO_4^{3-}
(apatite, monazite) and F^- (fluorite), an enhanced uptake of
americium was obtained in the pH range where the sorption is not
determined by hydrolysis.[3,17,27] For neptunium, a similar sorption
enhancement was observed for apatite, but not for fluorite. Some
sorption data (distribution coefficients) are given in Figure 5.
The magnitude of the sorption was higher than expected in comparison with data for low capacity minerals like quartz, corundum,
etc., when the cation exchange capacity is considered (see Table 3).
No enhancement of the sorption of Np(V) by F^- would be expected
since pentavalent actinide fluoride complexes are fairly soluble.

Cation Exchange Reactions. In Table 4, some measured sorption
data (distribution coefficients) are given for americium and
neptunium on high capacity minerals, and the cation and anion
exchange capacities for these minerals are given in Figure 6.
There is no apparent correlation between distribution coefficients
and exchange capacities for americium at pH 8 where americium
would be almost completely hydrolyzed. In the low pH range,
however, where Am^{3+} and $AmOH^{2+}$ would dominate, a very pronounced
salt dependence was observed for the sorption of americium on
montmorillonite (see Figure 7). This indicates that at least for
a highly charged unhydrolyzed ion like Am^{3+}, cation exchange may
be an important sorption mechanism (below the pH of hydrolysis).

Table 3. Cation exchange capacities for some low capacity minerals
          and distribution coefficients (Kd) for Am and Np.

Mineral	C.E.C.,[1] meq/kg	log Kd,[2] m^3/kg	
		Am	Np
Quartz	< 0.1	-0.1	-1.4
Fluorite	< 0.1	1.1	-2.2
Clacite	< 0.1	1.3	-0.5
Dolomite	< 0.1	1.3	
Apatite	0.1	1.2	0.8
Monazite	n.d.	1.5	

[1] For Na^+ at pH 8; particle size: 0.044 - 0.063 mm.

[2] At pH 8.

Table 4. Distribution coefficients (K_d) for Am(III) and Np(V) on
          high capacity minerals[d] (cf. Figure 8).

Mineral	Log K_d, m^3/kg			
	Am(III)		Np(V)	
	pH 5	pH 8	pH 5	pH 8
Montmorillonite	1.2	1.2	-1.0	-1.2
Muscovite	0.1	1.6		
Attapulgite	1.3	1.5	-0.8	1.2
Chlorite	1.2	1.5	-2.1	-1.0
Halloysite	0.8	1.2		
Kaolinite	0.3	1.3	-1.9	-1.0
Biotite	0.5	1.3	-1.6	-1.0
Serpentinite		1.9	-0.8	-0.4

[1] Artificial groundwater, ambient temperature, 7-12 g solid/L
water, particle size: 0.044-0.063 mm (except for montmorillonite); contact time: 5 d.

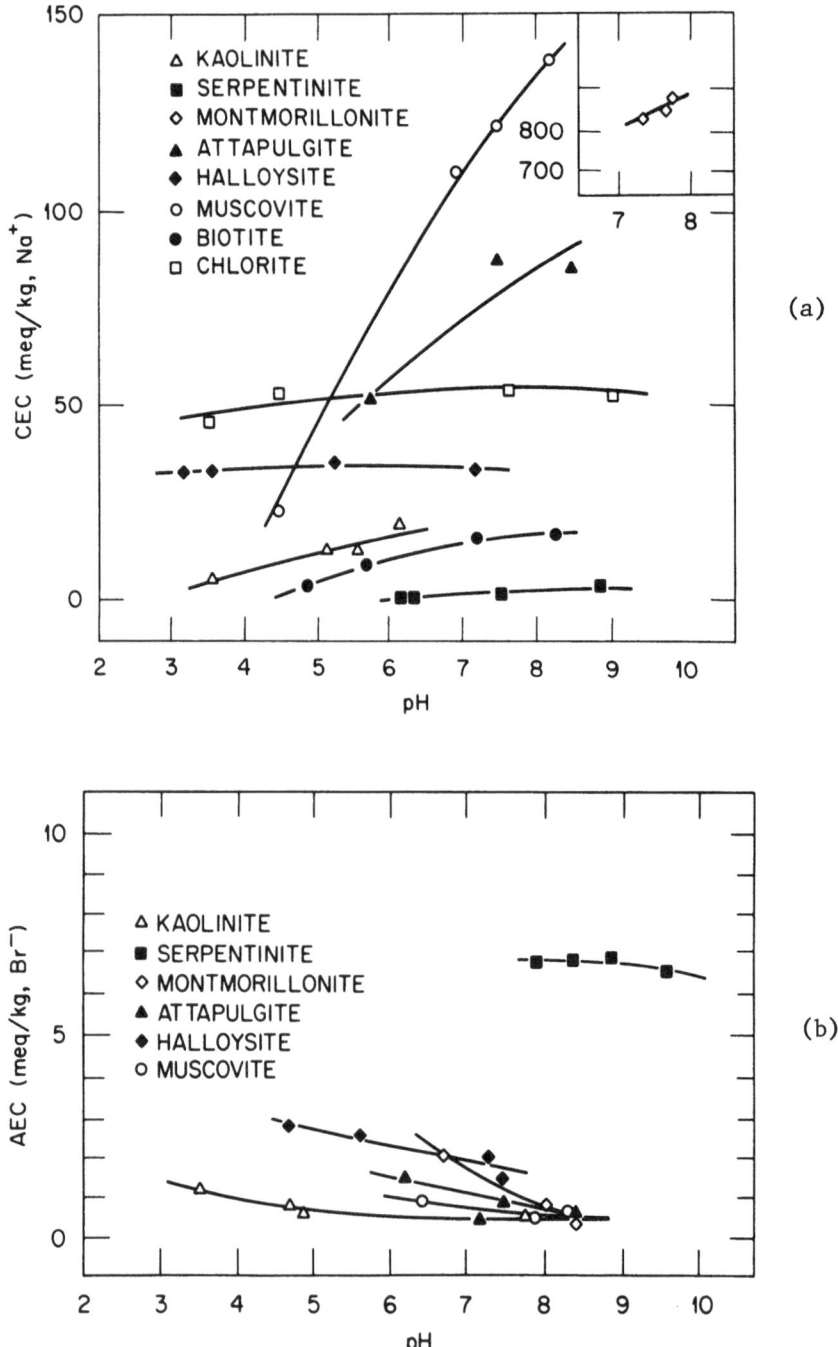

FIGURE 6: Cation and anion exchange capacities for some silicate
(a, b)   minerals.

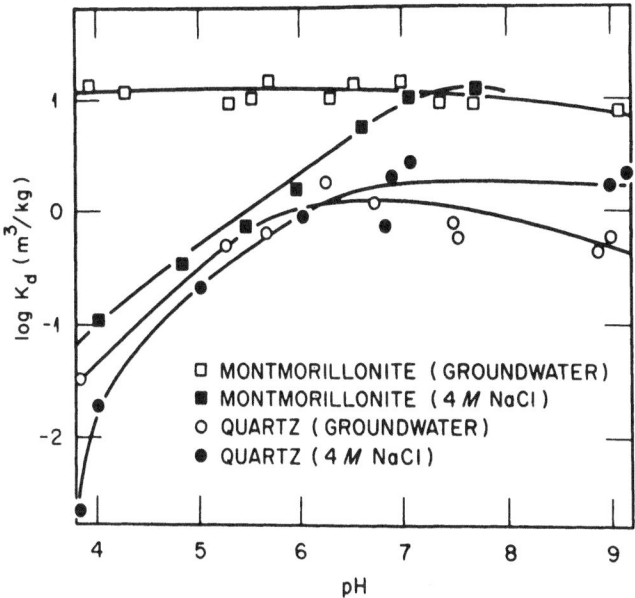

FIGURE 7: Sorption of Am(III) on montmorillonite and quartz in
groundwater and 4 M NaCl.

For Np(V) at pH 5, where largely NpO_2^+ would exist in solution,
there are no evident correlations between exchange capacities and
sorption although such correlations might be expected. Evidently
the large NpO_2^+ ion with two oxygen atoms along the axis and water
of hydration around the equatorial plane cannot easily occupy
exchange sites in the silicate minerals, possibly for steric
reasons.

Effects of Changes of Valence States. The sorption behavior
is entirely different for the various valence states, especially
due to the hydrolysis reactions (see above). A drastic change of
the sorption of, e.g., U(VI), Np(V) and Pu(IV) would be expected
by the reduction to the lower valence states, U(IV), Np(IV) and
Pu(IV), respectively, which would be expected to dominate in
certain anoxic natural waters (see above). Previously it has
been experimentally shown that U(VI) would be reduced to U(IV)[28]
and Np(V) to Np(IV),[29] by the presence of Fe(II)-containing
minerals in the system. This is further demonstrated in some
autoradiographic studies[30] (see Figure 8). The sorption of Np
under oxic conditions is concentrated on the Fe(II)-containing
minerals pyrite and biotite with a small background on quartz and
feldspars. The sorption on the pyrite and biotite is even stronger
in the absence of oxygen, indicating a possible reduction of
Np(V) to Np(IV) on the surface. For a surface that has been
oxidized with H_2O_2, a general nonspecific sorption is obtained.

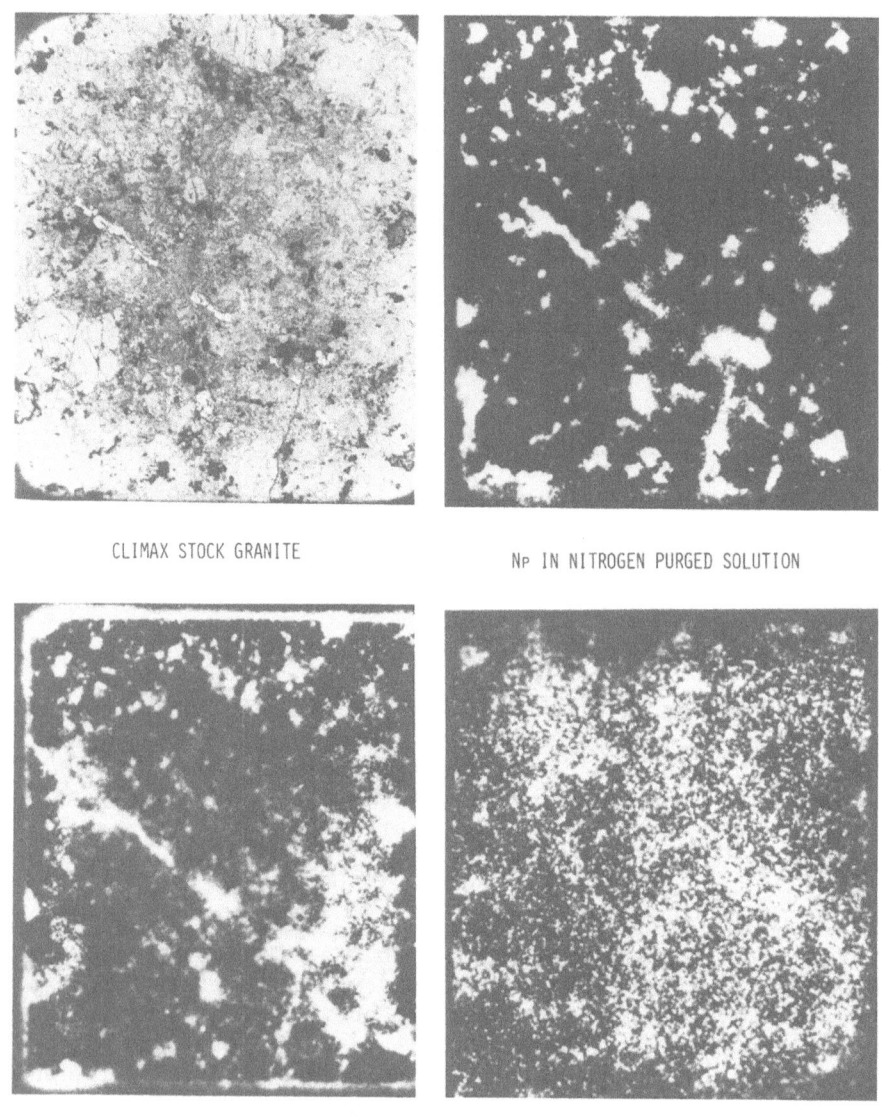

CLIMAX STOCK GRANITE

Np IN NITROGEN PURGED SOLUTION

Np IN AERATED SOLUTION

Np ON OXIDIZED SURFACE

FIGURE 8a: Autoradiograph of Np adsorption on climax stock granite under various oxidizing and reducing conditions.

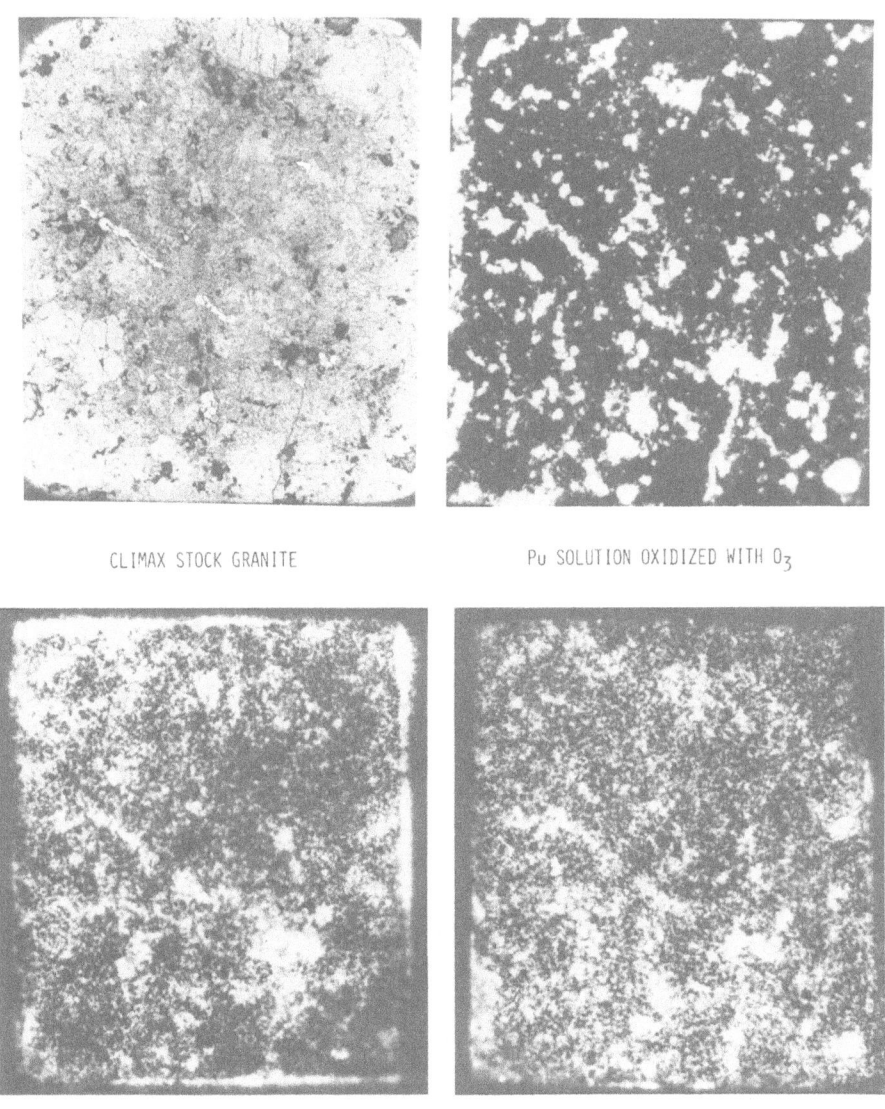

CLIMAX STOCK GRANITE   Pu SOLUTION OXIDIZED WITH O$_3$

Pu IN AERATED SOLUTION   Pu IN NITROGEN PURGED SOLUTION

FIGURE 8b: Autoradiograph of Pu adsorption on climax stock granite under various oxidizing and reducing conditions.

For plutonium, there is little difference between aerated and nitrogen-purged systems. The predominant pattern is a general sorption on the entire surface, which would be expected for Pu(IV). In a solution treated with O_3, where an oxidation to Pu(VI) has occurred, a sorption similar to the pattern for Np(V) is observed, that is, with a strong correlation with the Fe(II)-containing minerals. This, presumably, is evidence for reduction of Pu(VI) to Pu(IV) by Fe(II).

CONCLUSIONS

The most important chemical factors governing the sorption of actinides from aqueous solutions on geological media under environmental conditions are the reduction-oxidation potential of the system, determining the valence state, and pH, determining the degree of hydrolysis. Also of importance is the presence of potential complexing agents, either in solution or present in the solids. Under certain conditions (low pH, water composition) the exchange capacity of the components of the solid would be of importance for sorption. The simple cation exchange theory is, however, of no great significance in the environmental pH range.

ACKNOWLEDGEMENTS

This research was sponsored by the Office of Basic Energy Sciences, Division of Nuclear Sciences, U.S. Department of Energy, under contract (W-7705-eng-26) with the Union Carbide Corp. and by the Swedish National Council for Radioactive Waste Management.

REFERENCES

1. J.F. Relyea and R.J. Serne, "Controlled Sample Program Publication Number 2: Inter-Laboratory Comparison of Batch Kd Values", PNL-2872 UC-70, Pacific Northwest Laboratories, Richland (1979).
2. B. Allard, G.W. Beall and T. Krajewski, The sorption of actinides in igneous rocks, Nucl. Techn. 49(8):474 (1980).
3. G.W. Beall and B. Allard, Trans. Am. Nucl. Soc. Ann. Meeting 32:164 (1979).
4. B. Allard and G.W. Beall, work in progress.
5. C. Francis and D.F. Grigal, Soil Sci. 112:17 (1971).
6. C.M. Bower and J.O. Goertzen, Soil Sci. 87:289 (1959).
7. G. Jacks, "Groundwater at Depth in Granite and Gneisses", KBS Technical Report 88, Kärnbränslesäkerhet, Stockholm (1978).
8. M. Pourbaix, "Atlas of Electrochemical Equilibria", Pergamon Press, New York (1966).

9.   B. Allard, J. Rydberg, H. Kipitski and B. Torstenfält, Disposal of radioactive waste in granite bedrock, in: ACS Symp. Ser. 100, American Chem. Soc., Washington, D.C. (1979), 47.

10. Y. Tarsy and R.M. Garrels, Geochim. Cosmochim. Acta 38:1101 (1974).

11. I. Grenthe, Determinations of redox potentials in groundwater from Stripa and Finnsjön, App. 5 in: "Copper as Encapsulation Material for Unreprocessed Nuclear Fuel Work", KBS Technical Report 90, Kärnbränslesäkerhet, Stockholm (1978).

12. S. Ahrland, "Structure and Bonding, Volume 5", Springer-Verlag, Berlin (1978).

13. S. Ahrland, J.O. Liljenzin and J. Rydberg, Actinide solution Chemistry, in: "Comprehensive Inorganic Chemistry, Volume 5", Pergamon Press, Oxford (1973).

14. B. Allard, H. Kipatsi and J.O. Liljenzin, J. Inorg. Nucl. Chem. (1980), in press.

15. B. Allard, unpublished data.

16. T. Paces, Geochim. Cosmochim. Acta 33:591 (1969).

17. G.L. Johnson and L.M. Toth, "Plutonium (IV) and Thorium (IV) Hydrous Polymer Chemistry", ORNL/RM-6365, Oak Ridge National Laboratory, Oak Ridge, TN (1978).

18. L.G. Sillen, "Stability Constants of Metal-Ion Complexes", Chem. Soc. Spec. Publ. No. 17 and 24, The Chemical Society, London (1964, 1971).

19. C.F. Baes and R.E. Mesmer, "The Hydrolysis of Cations", John Wiley and Sons, Toronto (1976).

20. B. Allard and G.W. Beall, work in progress.

21. I.E. Staric, "Grundlagen der Radiochemie", Akademie-Verlag, Berlin (1963).

22. F. Kepak, Chem. Rev. 71:357 (1971).

23. M. Haissinsky, "La chimie nucléaire et ses applications", Masson, Paris (1957).

24. E. Matijevic, M.B. Abramson, R.H. Ottewill, K.F. Schultz and M. Kerker, J. Phys. Chem. 65:1724 (1961).

25. P.H. Tewari and W. Lee, J. Colloid Interface Sci. 52:77 (1975).

26. R.O. James and T.W. Healy, J. Colloid Interface Sci. 40:43,53,65 (1972).

27. T. Krajewski, G.W. Beall, B. Allard and J. Peterson, Trans. Am. Nucl. Soc. Ann. Meeting 32:168 (1979).

28. B. Allard, H. Kipatsi, B. ans Torstenfält, "Sorption of Long-Lived Radionuclides in Clay and Rock. Part II", KBS Technical Report 98, Kärnbränslesäkerhet, Stockholm (1978).

29. E. Bondietti, private communication.

30. G.W. Beall, D. O'Kelley and B. Allard, "An Autoradiographic Study of Actinide Sorption on Climax Stock Granite", ORNL report #5617, Oak Ridge National Laboratories, Oak Ridge, TN (1980).

31. P. Fritz, J.F. Barker and J.E. Gale, "Geochemistry and
     Isotope Hydrology of Groundwaters in the Stripa Granite",
     LBL-8285, Lawrence Livermore Laboratory, Berkeley (1979).
32. R.M. Smith and A.E. Martell, "Critical Stability Constants,
     Volume 4: Inorganic Complexes", Plenum Press, New York
     (1976).
33. R.J. Lemire and P.R. Tremaine, "Uranium and Plutonium
     Equilibria in Aqueous Solution to 200°C", AECL-6655, Atomic
     Energy of Canada Limited, in press.
34. G. Choppin, Trans. Am. Nucl. Soc. Ann. Meeting 32:166 (1979).

SURFACE CHEMISTRY AND ADSORPTION PROPERTIES

OF MILLED CHRYSOTILE ASBESTOS FIBRES

Syed M. Ahmed

Physical Chemistry Section
Mineral Sciences Laboratories
CANMET, Energy, Mines and Resources Canada, Ottawa

ABSTRACT

The chrysotile-aqueous solution interface has been examined
by measuring the zpc, the surface charge densities and the zeta
potential, by means of potentiometric-pH titrations and electrophoresis. The zpc of these samples after cleaning by dilute acid
(pH 3) and washing with water, occurs at pH 10.2, but shifts to
higher pH values if the dissolved Mg^{2+} is adsorbed on the surfaces.
The surface is positively charged below the zpc and the apparent
charge densities, when plotted against pH, show two waves. The
first wave between the zpc and pH 6 is due to the reversible
double layer formed by the basic dissociation of the surface
hydroxyl groups and or adsorption of H^+. The second wave is due
to the displacement of the structural surface-Mg^+ by H^+ which
eventually leads to an irreversible disintegration of the structure and loss of mechanical properties. These conclusions are
supported by a thermodynamic analysis. Silicate ions are easily
adsorbed on chrysotile shifting the zpc to a lower pH; they also
act as anionic activators for the adsorption of long-chain cationic
amines. Silicate ions can also protect the surface from acid
damages in slightly acid solutions.

INTRODUCTION

A knowledge of the surface chemistry of asbestos and its
application in wet processes such as separation, sedimentation
and filtration has proved very useful in improving efficiency and
quality control in the asbestos industry. At present wet methods
for processing of asbestos are used only to a limited extent.

Because of the health hazard involved in dry handling of asbestos,[1] wet methods for complete processing would definitely be preferable. However, wet methods also give rise to certain characteristic problems which can hopefully be solved, but only after achieving a full understanding of the surface chemistry of asbestos. A serious problem with the chrysotile fibre, for example, is the disintegration of its mechanical, physical and structural properties in acid solutions. Improper use of surfactants may cause serious adverse effects such as altering the wettability of the fibre and the bonding strength to the cement in asbestos-reinforced concrete, foaming in the pulp, and fibre floating on the surface during wet processing. Fibre damage, if any, in the wet handling should be minimized and wet grading methods and standards have also to be developed. Methods should be developed to remove all traces of asbestos suspensions in effluent water before discharging.

For establishing optimum conditions for wet processing methods, the surface chemistry of all forms of asbestos should be understood thoroughly. The following work on the chrysotile-solution interface was undertaken as part of a study by the Industrial Minerals Laboratory of CANMET in developing all-wet processing methods for asbestos. The subject has been reviewed in another publication.[2]

EXPERIMENTAL PROCEDURE

Processed chrysotile fibre from various Canadian sources, Carey Canadian Ltd. (7RF), Lake Asbestos (7R) and Canadian Johns' Manville (CJM) (7RS), were supplied by the Mineral Processing Laboratory of CANMET, Ottawa. The designation, e.g., 7RF, refers to the grade as described elsewhere;[2,3] the first number refers to the fibre length (7 is the shortest), and the first letter, R, refers to the degree of fibre opening, while F refers to float (fines). From X-ray analysis, electron microscopy and thermogravimetric analysis (TGA) studies, the samples were found to show high degree of fibre opening and contained minor amounts of magnetite, brucite (0.6%) and platy serpentine, the major constituent being chrysotile fibre. This particular sample of asbestos from Carey Canadian contained low amounts of brucite, although about 5% of brucite has been reported in a recent analysis of other samples.[4] While magnetite and platy serpentine would not affect the present studies, the brucite content was small and was further reduced by a surface cleaning process as described later. The cleaned samples were further dispersed in salt solutions and the suspensions were used for electrophoretic measurements after adjusting the pH as desired.

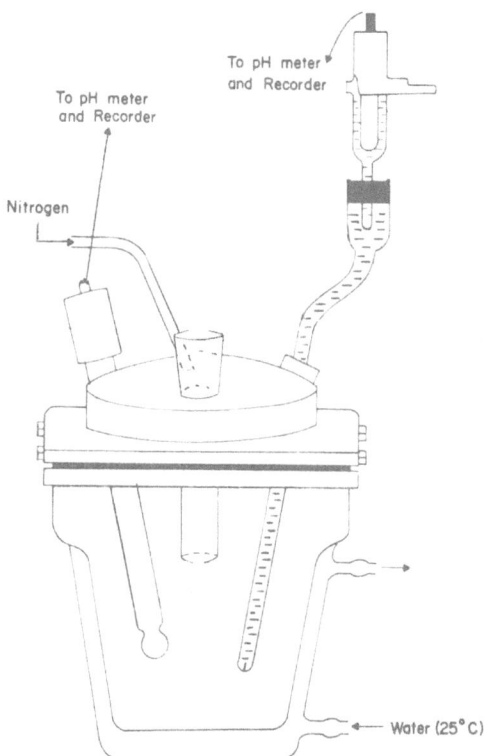

FIGURE 1   Reaction vessel for potentiometric pH titrations.

A Zeta Meter was used for measuring electrophoretic mobility using standard procedure and precautions. However, the original electrodes of the Zeta Meter were replaced by palladium electrodes which were partly charged with electrolytic hydrogen as described elsewhere.[5,6]

The surface charge densities and the zero point of charge (zpc)* of chrysotile fibre were obtained from potentiometric measurements of the pH changes in KNO_3 solutions of constant ionic strength, caused by the addition of known amounts of asbestos. The asbestos was freshly washed with distilled water on a filter paper on a Buchner funnel under suction and rinsed and dried before weighing and transferring to the reaction vessel. . Nitrogen was bubbled through the solution initially to remove CO_2, following which a stream of nitrogen was maintained above the solution. A Metrohm flask with an outer water jacket (Figure 1) was used for the pH titrations while the water at constant

---

*the pH at which surface has net zero charge.

temperature was circulated through the water jacket. Hydrochloric
acid and potassium hydroxide solutions were used to adjust pH.
The pH readings before and after adding the asbestos were recorded
as a function of time, using a Beckman Research pH meter and a
strip chart recorder. From the predetermined values of experimental activity coefficients of H^+ and OH^- in the KNO_3 solutions
of known ionic strength, and the recorded values of ΔpH (or ΔpOH
above pH 7), the charge densities were calculated. A surface
area of 100 m^2/g was assumed to bring these surface charge densities to the expected values and this is in agreement with the
published data on surface areas of highly opened and broken
fibre.[7] The procedure has been described in detail elsewhere.[8,9]
A typical run is shown in Figure 2, where the final pH occurs in
the alkaline range and almost a constant equilibrium pH is reached
after about 15 minutes. This fast, initial pH change during the
first 15 minutes is attributed to the surface dissociation and
adsorption process, while the later, slower change of pH with
time is due to slow solubility effects. However, this distinction
between these two stages of interfacial reactions based on kinetic
considerations was not so well defined when the final pH after
reaction occurred in the acid medium (< 5). In such cases the
reaction continued as long as a half hour, until the pH variations
with time became constant and negligibly small. As seen later,
such behavior occurred when H^+ started to replace Mg^{2+} from the
surface.

RESULTS

     The variation with pH of the apparent charge density in
$\mu C/cm^2$, as obtained from the Δa_{H^+} (or Δa_{OH^-}) changes for the

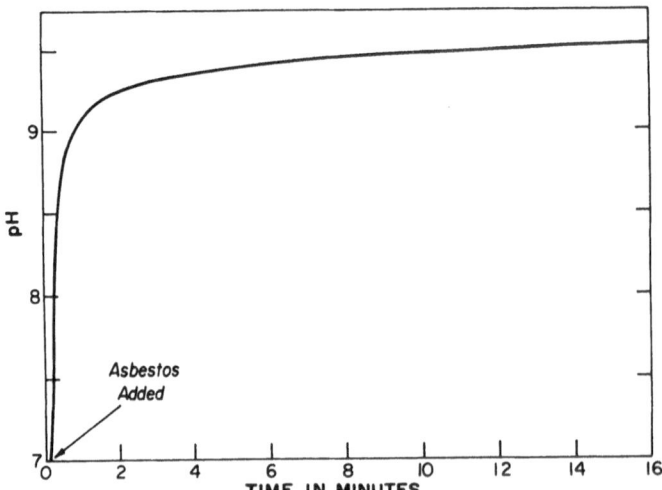

FIGURE 2   Variation of pH with time before and after adding the
           chrysotile sample to the solution.

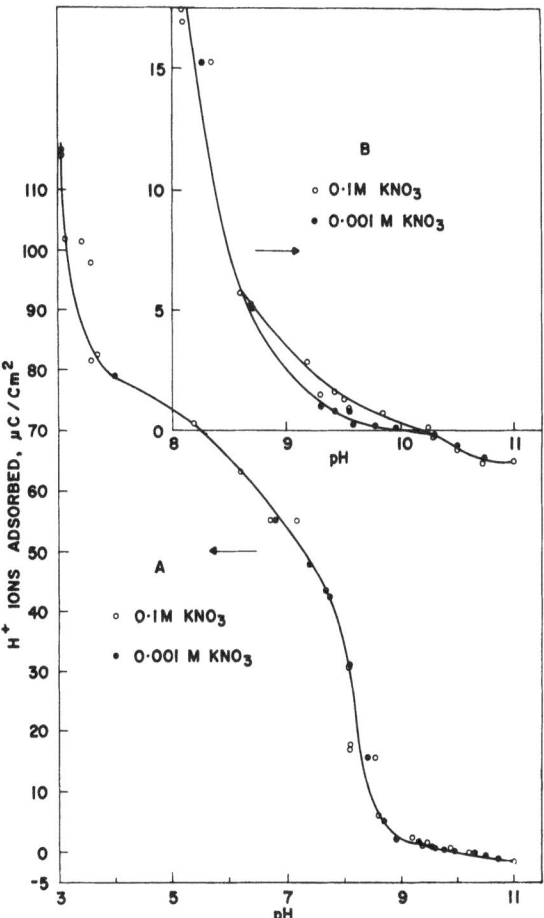

FIGURE 3   Variation of the apparent surface charge density of
           chrysotile (Carey Canadian 7RF) with pH in 0.001 M and
           0.1 M KNO_3 solutions.

reaction, is shown in Figure 3 for the pH range 3-11 for 0.001 M
and 0.1 M KNO_3 solutions.   The same results for the pH range
8-11 are shown on an expanded scale in Figure 3B.

    The electrophoretic mobilities, after converting into the
zeta potential, using standard equations[10] have been plotted
against pH, for different reagents and experimental conditions,
in Figures 4 to 8.   Corrections to the particle mobility or zeta
potential for the particle shape have not been applied.   Only
relative variations and not absolute values of the zeta potentials
in these figures are important for the present discussion.   All
chrysotile samples tested and shown in Figure 4 show a positive
zeta potential and hence positive surface charge with no reversal
of sign.   A similar behavior has been reported by Edwards et al.

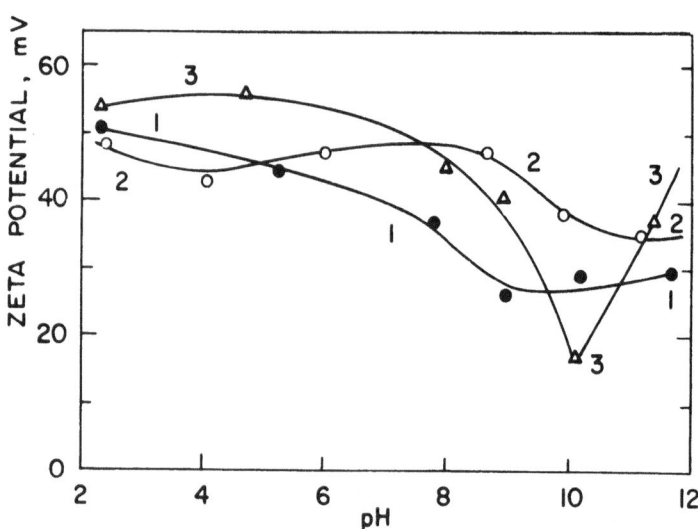

FIGURE 4   Variation of the zeta potential with pH as measured
from the electrophoretic mobility of three different
Canadian chrysotile samples in 1×10^{-3} M KNO_3.
1. Lake Asbestos 7R;   2.   Carey Canadian 7RF;   3.
CJM-7RS.

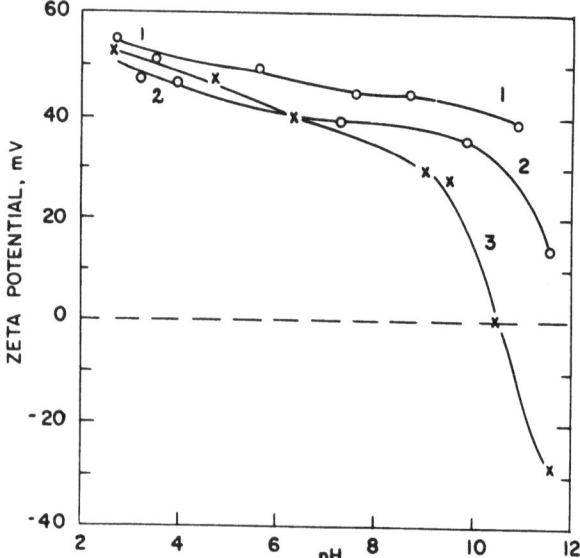

FIGURE 5   Variation of zeta potential of chrysotile (CJM-7RS)
suspensions with pH: 1. in 1×10^{-3} M KNO_3;
2. in 1×10^{-2} M KNO_3,   and   3. in 1×10^{-2} M KNO_3 after
surface cleaning by washing with a dilute acid, and
after complete removal of the acid.

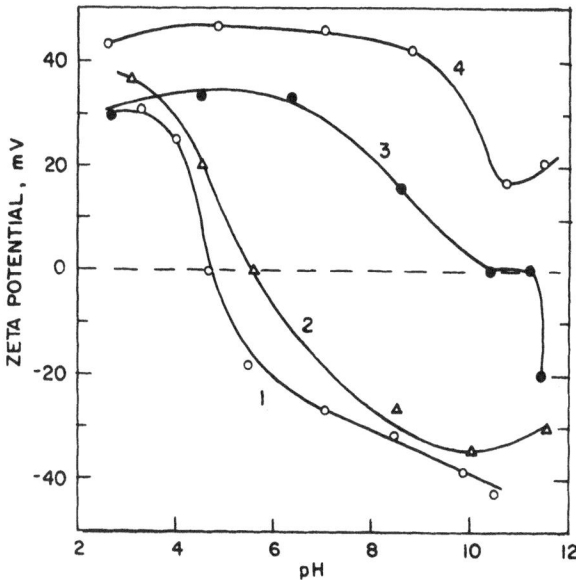

FIGURE 6   Effect of sodium silicate on the zpc and zeta potential
of chrysotile (CJM-7RS) in 1×10^{-2} M KNO_3 containing:
1. 1×10^{-3} M silicate conditioned for two hours;
2. 1×10^{-3} M silicate with pH adjusted just before
measurement;   3. 5×10^{-4} M silicate;   4. 1×10^{-5} M silicate.

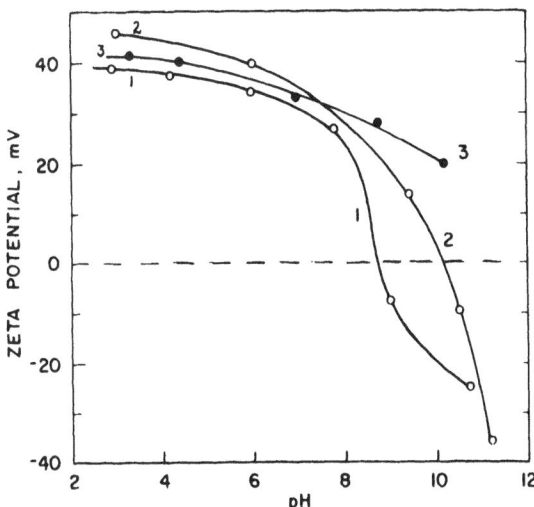

FIGURE 7   Effect of dodecylamine (DDA) on the zeta potential of
chrysotile (CJM-7RS), treated with sodium silicate in
1×10^{-3} M KNO_3 solutions: 1. 1×10^{-5} M DDA + 1×10^{-3} M
silicate;   2. 1×10^{-4} M DDA + 1×10^{-3} M silicate;
3. 1×10^{-3} M DDA + 1×10^{-3} M silicate.

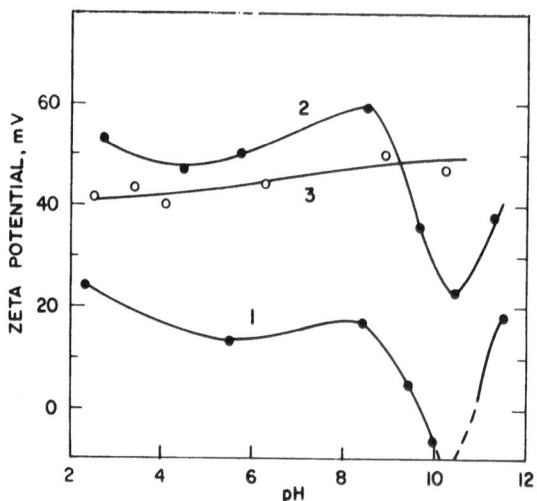

FIGURE 8   Variation of the zeta potential of chrysotile (CJM-7RS)
            with pH in: 1. 1×10^{-2} M KNO_3 + 1×10^{-3} M SO_4^{2-} (after
            4 h of contact); 2. 1×10^{-3} M KNO_3 + 2.3×10^{-4} M DDA;
            3. 1×10^{-3} M KNO_3 + 1×10^{-3} M lauryl sulphate.

recently.[11]   However, if the sample is washed with a dilute HCl
(pH 3) for a few seconds and excess acid washed out before com-
mencing the electrophoretic experiment, then a reversal of charge
is noticed with a zpc occurring at pH 10.2, as shown in Figure 5.
The reversal of surface charge and the shift in the zpc to lower
pH values in Na_2SiO_3 solutions, caused by the adsorption of
silicate ions, is shown in Figure 6. This behavior is very
similar to what Martinez and Zucker have reported earlier.[12]
However, the adsorbed silicate ion could act as an anionic activator for the adsorption of dodecylamine (cationic), as is shown
in Figure 7.   In Figure 8, the effects of SO_4^{2-}, lauryl sulphate
and dodecylamine on the zeta potential are shown.

DISCUSSION

Surface Structure of Chrysotile

     Chrysotile is a hydrated, monoclinic form of magnesium
silicate with a general formula $Mg_3Si_2O_5(OH)_4$ or $3MgO \cdot 2SiO_2 \cdot 2H_2O$,
containing small amounts of Fe, Cr, Co, Ni and Sc as impurities.
The fibre is made of units containing an inner layer of silicon-
oxygen tetrahedra with an outer, octahedra of MgOH;   several such
layers form a tubular structure.[13,14]   The curvature and hence
the tubular structure have been attributed to structural misfit
between the brucite and silica groups,[15,16] shown schematically
in Figure 9A.   The SiO_4 tetrahedra are joined to the brucite

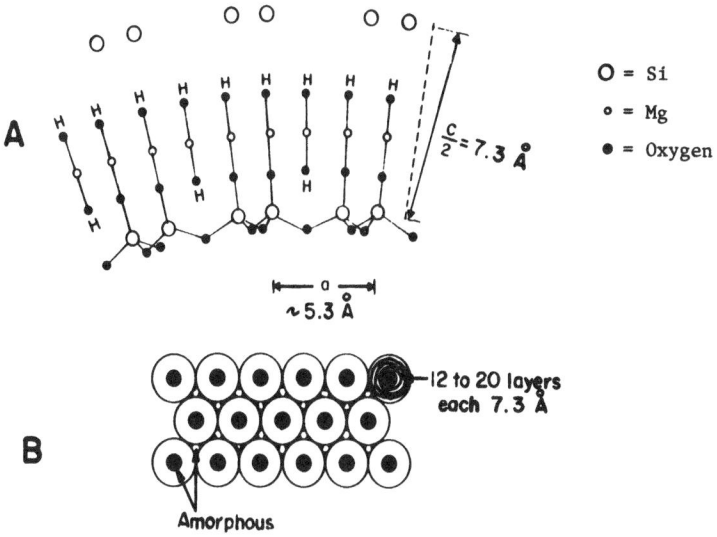

FIGURE 9 A. Surface structure of chrysotile fibre,
          B. Cross section of chrysotile fibres in the mineral
          based on References (12) and (14).

layer through oxygen and in between every two $\geq$Si–O–MgOH layers
is an independent brucite layer, all held together in a strong,
hydrogen bonded structure. The fibre walls are believed to
consist of 12 to 20 such layers, each one being about 7.3 Å in
thickness. A cross-sectional view of the stack of cylindrical
fibrils is shown in Figure 9B[17] each cylinder having an outer
diameter of 200–250 Å and an effective inner diameter of 20–50 Å.
Atkinson et al.[18] have reported 300–375 Å as the fibril diameter
and 1.4–1.8 mm as fibril length for Canadian chrysotile samples
from high resolution electron microscopy. The pore volume is as
high as 30% of the fibre volume[17] and the inter- and intra-
fibrillar space has been found to be filled with an amorphous
form of magnesium silicate having the same composition as chrysotile. The most interesting surface characteristic of chrysotile
fibre is that there are two kinds of Mg–OH surface groups.
The removal of –MgOH groups chemically for example, will expose
the hydrated $\geq$Si–O$^-$ groups at the outer surface. As a result,
the fibre is highly susceptible to attack by acids.[19-22] Removal
of Mg^{2+} results in a less hydrated structure and a breaking of
the hydrogen bonds that give mechanical strength to the fibre. A
gradual unrolling of the fibre has been found to occur giving
rise to ribbons, and the fibre thus loses its mechanical strength,[19]
the change being directly proportional to the extent of Mg^{2+}
removal.

A detailed understanding of the chrysotile–solution interfacial reactions is important in all wet processes.

Interfacial Reactions and Surface Charge

The behavior of the chrysotile-solution interface, from a surface-charge point of view, is similar to that of amphoteric oxides that have been studied earlier.[8,9] However, chrysotile has a rather high zpc (10.2) and two distinct equilibria can be distinguished below the zpc, at the chrysotile-electrolyte interface (Figure 3). The interfacial reactions are as shown below:

pH 10 (zpc)   pH > 10

$$Si-O-MgOH \underset{H^+}{\overset{OH^-}{\rightleftharpoons}} Si-O-MgO^- \qquad (1)$$

$$OH^- \uparrow\downarrow H^+ \qquad\qquad\qquad\qquad (2.A)$$

$$\left.\begin{array}{l} Si-O-Mg^+ \\[6pt] or \\[6pt] Si-O-MgOH_2^+ \end{array}\right\} \xrightarrow{H^+} Si-OH \qquad (2.B)$$

pH ∿ 5.5–10   pH 2

An acidic dissociation of the surface-MgOH groups (Equation (1)) in the pH range > zpc results in a decrease of pH on adding the crysotile to the solution and hence the negative surface charge (Figure 3).

The two waves observed in the surface charge vs. pH plot in Figure 3 are attributed to the two stages of surface reactions that take place below the zpc, as shown in Equation (2), equilibria A and B.

In the first stage (A), as the pH is decreased from 10 (zpc) to about 6 (Figure 3B), an acidic dissociation of the $-Mg(OH)$ groups, and/or H^+ adsorption occurs on the surface, which accounts for the positive surface charge and a positive zeta potential, and an increase in pH when chrysotile is added to the electrolyte solutions. The NO_3^- (counter ions) in the present case do not replace the primary hydration layer (Helmholtz layer). Hence the surface charge densities (q) in Figure 3B are practically independent of NO_3^- concentration, except for the small increase in q^+

due to compression of the double layer, as the ionic strength is increased from 0.001 M to 0.1 M KNO_3. This behavior is typical of amphoteric oxide surfaces which do not show any tendency for the specific adsorption of anions.[8,9]

As the pH is decreased further below about 5, a replacement of the structural $-Mg^+$ by H^+ starts to occur as a second step (Equation (2.B)), resulting in a rapid increase in the H^+ consumption from the solution and consequently an increase in pH. Hence the q^+ values, quoted in Figure 3, below pH 5, are not real, because this is an ion exchange process where a divalent Mg^{2+} ion is replaced by H^+, resulting in $\geqslant Si-OH$ groups on the surface and a disintegration of the chrysotile structure, as described previously.

As step A tends to increase the surface charge while step B tends to decrease it, a kinetic equilibrium should be reached between the two steps whereby the real surface charge density and zeta potential should level off in the acid medium (see Figures 4-7) until all the surface $-Mg^+$ sites are replaced by H^+, when a second isoelectric point, characteristic of the silica groups, is reached at pH 2. The zeta potential falling towards zero at pH 2 has been reported by Martinez and Zucker.[12] However, such decrease in the electrophoretic mobility at pH 2 is not recorded in Figures 4-7, because of the kinetically slow removal of Mg^{2+}, as stated above. The zeta potential of chrysotile has also been measured by Leight and Wei more recently.[23]

Surface Reversibility and Stability

Because of the formation of a reversible, electrical double layer, the interface reactions up to about pH 6 are also reversible, particularly in the presence of some added Mg^{2+} in solution. However, once the pH is reduced further and Mg^{2+} ions displaced, an irreversible, physical and structural damage begins to occur and the regeneration of the same surface will not be possible, although the damage to the inner fibrils will be slow and time dependent, as H^+ ions have to diffuse through the outer $\geqslant Si-OH$ groups. Further details are discussed in a separate review paper.[2]

The pH limits for the stability of the hydrated magnesium silicate can also be established from the thermodynamics of the overall equilibria written as:

$$Mg_3Si_2O_5(OH)_4 + 6H^+ = 2Si(OH)_4 + 3Mg^{2+} + H_2O, \qquad (3)$$

for which, $K = \dfrac{\left[Si(OH)_4\right]^2 \left[Mg^{2+}\right]^3}{\left[H^+\right]^6}$ (4)

or $\log K = 2 \log \left[Si(OH)_4\right] + 3 \log \left[Mg^{2+}\right] + 6pH$ (5)

From the free energy of formation of the hydrated magnesium silicate, log K is found to be 34.0,[24] so that for the $Si(OH)_4$ concentrations of $10^{-2.6}$ mol L^{-1} (saturation solubility) and 10^{-12} mol L^{-1}, the solubility or the stability vs. pH lines are shown by lines A and B, respectively, in Figure 10. An increase in the silicate ion concentration from 10^{-12} mol L^{-1} to the saturation level is shown to extend the stability limits of the fibre to a lower pH of 6 (minimum limit) by the common ion effect (Figure 10). It might, similarly, be possible to further extend the stability limits to lower pH levels by using other anionic complexing agents or surfactants.

Figure 10 predicts that the fibre is unstable below about pH 6.0 unless Mg^{2+} concentration in solution is 1 M (or 0.1 M at pH 7) which is in agreement with the surface chemical behavior shown in Figure 3. In acid solutions, dissolution of Mg^{2+} and a consequent increase in pH occurs until the stability conditions (Figure 10) are reached, which would prevent further dissolution.

Effect of Activators and Surfactants

In solutions saturated with Mg^{2+}, a self activation of the negatively charged, chrysotile surface occurs by adsorption of

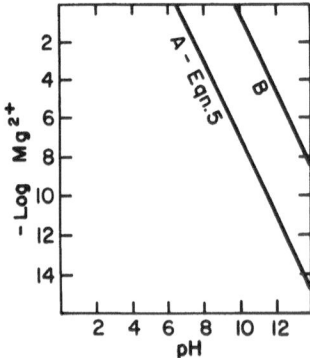

FIGURE 10   Stability diagram for hydrated magnesium silicate showing: A. saturation concentration of $Si(OH)_4$ solution = $10^{-2.6}$ mol L^{-1} and B. $Si(OH)_4$ concentration = 10^{-12} mol L^{-1}. Based on Reference (20).

the dissolved Mg^{2+} forming a secondary brucite layer which has a zpc close to pH 12. Hence, the measured zpc may vary from a pH 10 for a purely chrysotile surface as found in this work and Pundsack's work[25] to a pH = 11.8 for a purely brucite surface as found by Martinez in streaming potential measurements, where the solution can get easily saturated with Mg^{2+}. Addition of silicate ions was shown to shift the zpc to pH 6; silicate ions would also shift the pH of minimum solubility. Also, the silicate, when adsorbed on the positively charged chrysotile below about pH 10, can act as an anionic activator in the adsorption of a long chain cationic amine as shown in Figure 7. Such adsorption would not only stabilize the surface but also make it hydrophobic. Lauryl sulphate may also be expected to be adsorbed on the positively charged surfaces of chrysotile and to reduce the zeta potential but for some reason, this does not seem to take place as seen in Figure 8. The SO_4^{2-} ions also have very little effect on the zeta potential (Figure 8) which implies that SO_4^{2-} is not specifically adsorbed on the $-MgOH_2^+$ surface of chrysotile.

The effects, if any, of adding silicate and organic surfactants on the acid attack and fibre disintegration need to be investigated in detail for further development of the wet processing methods.

In conclusion, the zpc of cleaned chrysotile fibre occurs at pH 10.2 which is also the pH of the maximum settling rate.[25] However, in the presence of excess dissolved Mg^{2+}, the zpc shifts to a higher pH. The surface reactions in the pH range below the zpc occur in two stages. A reversible, basic dissociation of the $-MgOH$ groups occurs down to pH 6 followed by an irreversible Mg^{2+} replacement in more acidic solutions causing slow structural damage. The latter reaction can be prevented by increasing the Mg^{2+} concentration substantially (Figure 10). However, the structural damage is small for short durations of acid treatment and the surface can be stabilized by adding silicate ions, care being taken that the presence of the surface Mg^{2+} is necessary for good bonding properties of asbestos with cement. Silicate ions in addition to shifting the zpc to lower pH values, can also act as anionic activators for dodecylammonium ion adsorption.

Concerning the slow lung damage that the asbestos fibre is known to cause, one should investigate the following as a possible mechanism. The fibre once lodged into the cell membrane could undergo slow leaching of Mg^{2+} in the slightly acid environment, and get converted into hydrated silica thereby blocking the pores by swelling. This could seriously interfere with the ion permeability, exchange and other transport properties of the membrane, initiating other biological complications.

ACKNOWLEDGEMENTS

Potentiometric titrations were performed by M. Pollitt and electrophoretic measurements were carried out by J. Margeson.

REFERENCES

1. I. Webster, "Proceedings of Asbestos Symposium, Johannesburg, South Africa, 1977 October 3-7", H.W. Glen, ed., National Institute of Metallurgy, Randburg (1978). p. 3.
2. S.M. Ahmed and A.A. Winer, "Chrysotile Asbestos, A Review", CANMET/MSL Publication, 1980. In preparation.
3. "Manual of Testing Procedure for Chrysotile Asbestos Fibre, 111th Ed., Revised 1978" (Canada), Asbestos Textile Inst. Inc. and Quebec Asbestos Mining Assocn.
4. M. Cossette, A.A. Winer and R. Steele, "Fourth International Conference on Asbestos, Torino, 1980 May", Paper II.A-B/1, p. 201.
5. R. Neihof, J. Colloid Interface Sci. 30:128 (1969).
6. S.M. Ahmed, "Mines Branch Technical Bulletin, TB 140", Department of Energy, Mines and Resources, Ottawa, Canada (1971).
7. E. Papirer, P. Roland and J.B. Donnet, "Fourth International Conference on Asbestos, Torino, 1980 May", Paper II.A-B/4, p. 249 (see Figure 8).
8. S.M. Ahmed, "The Anodic Behavior of Metals and Semiconductors Series, Vol. 1, Oxides and Oxide Films", J.W. Diggle, ed., Marcel Dekker, New York (1972). pp. 319-517.
9. S.M. Ahmed, J. Phys. Chem. 73:3546 (1969).
10. D.J. Shaw, "Electrophoresis", Academic Press, London (1969).
11. C.R. Edwards, W.B. Kipkie and G.E. Agar, International Journal of Mineral Processing 7:33 (1980).
12. E. Martinez and G.L. Zucker, J. Phys. Chem. 64:924 (1960).
13. E. Martinez and J.J. Comer, Am. Mineralogist 49:153 (1964).
14. C.W. Daykin, "The Physics and Chemistry of Asbestos Minerals, Second International Conference, 1971 September", Louvain Univ., Belgium, Paper 2:6.
15. E.J.W. Whittaker, Acta Cryst. 10:149 (1957).
16. E. Martinez, Trans. C.I.M. LXIX:414 (1966).
17. F.L. Pundsack, J. Phys. Chem. 65:30 (1961).
18. A.W. Atkinson, R.B. Gettins and A.L. Rickards, in: "Physics and Chemistry of Asbestos Minerals", Second International Conference, 1971 September, Louvain Univ., Paper 2:4.
19. A.W. Atkinson and A.L. Rickards, ibid, Paper 3:1.
20. J. Goni, Z. Johan and C. Sarcia, ibid, communication.
21. A. Morgan, A. Holmes and A.E. Lally, ibid, Paper 2:8.
22. L.J. Monkman, ibid, Paper 3:2.
23. G. Light and E.T. Wei, Environmental Research 13:135 (1977).
24. W.E. Wildman, L.D. Whittig and M.L. Jackson, Am. Mineralogist 56:587 (1971).
25. F.L. Pundsack, J. Phys. Chem. 59:892 (1955).

X-RAY PHOTOELECTRON SPECTROSCOPIC (XPS) STUDIES ON THE CHEMICAL NATURE OF METAL IONS ADSORBED ON CLAYS AND MINERALS

J.G. Dillard, M.H. Koppelman, D.L. Crowther
and C.V. Schenck
Department of Chemistry
Virginia Polytechnic Institute and State University
Blacksburg, Virginia 24061

J.W. Murray and L. Balistrieri
Department of Oceanography
University of Washington
Seattle, Washington 98195

ABSTRACT

Adsorption of selected metal ions including Pb(II) and Co(II) on mineral and clay substrates has been carried out. The chemical nature of the adsorbed metal ion has been examined using X-ray photoelectron spectroscopy (XPS). The interpretation of the XPS results indicates that lead (II) is oxidized to lead (IV) on β- and δ-MnO_2, and Co(II) is present as Co(II) on illite. The oxidizing agent for Pb(II) is believed to be O_2 or MnO_2 and a reaction process consistent with the XPS results has been proposed.

INTRODUCTION

The interaction of metal ions with minerals has been the subject of intensive studies in the areas of clay mineralogy[1,2] and in marine chemistry.[3] Although many investigations have established the extent of metal ion uptake, limited information is available on the chemical nature (oxidation state, coordination geometry) of the adsorbed metal ion.[4-10] Data on surface speciation has been obtained through reflectance UV-vis, infrared spectroscopy,[10-13] and XPS studies.[4-9,14,15]

In an XPS study of the adsorption of lead (II) on montmorillonite,[4] binding energy measurements were used to establish the oxidation state of adsorbed lead. From a comparison of the binding energies for Pb, PbO, and PbO_2, with that for adsorbed Pb, it was concluded that the electronic environment about adsorbed lead is like that for PbO.[4]

The adsorption of Co(II) on δ-MnO_2 has been found to proceed with oxidation of Co(II) to Co(III).[9] The presence of Co(III) was established from XPS measurements.[9] In the XPS analysis of cobalt (II) on Al_2O_3 and on ZrO_2,[5,6] the comparison of binding energies for adsorbed cobalt (II) with those for authentic cobalt oxides and hydroxides revealed that cobalt adsorbed on Al_2O_3 and on ZrO_2 exists as $Co(OH)_2$. In the XPS study of cobalt adsorption on zeolites, Lunsford et al.[16] indicate that $Co(H_2O)_6^{2+}$ is adsorbed as Co(II) and $Co(NH_3)_6^{3+}$ is adsorbed as Co(III) on Y-type zeolites. Analysis of XPS spectra for Co(II) adsorbed on clay minerals as a function of pH was interpreted to suggest an aqua Co(II) species.[7] No evidence for the formation of cobalt hydroxide was found.

The mechanism for the reduction of iron in nontronite has been probed[8,17] using XPS. It is found that the reduction process proceeds via two steps. In the first step, structural Fe(III) is reduced to Fe(II) and no structural changes occur. In the second step, additional Fe(III) is reduced, structural OH groups are lost, and the coordination number about Fe(II) in the octahedral layer is lowered.

The adsorption of Ba(II) on calcite has been studied.[18] Comparison of atomic absorption and XPS methods for quantitative analysis of Ba(II) suggested that Ba(II) could be detected at trace levels, $\simeq 10^{-8}$ g. The results of the combined study revealed that Ba(II) adsorption is dependent on the surface area of calcite and on the initial Ba(II) concentration.

XPS has been employed to study the dissolution of feldspars.[19] From measurements of potassium, aluminum, and silicon peak intensities, it is concluded that feldspar dissolution is controlled by processes at the solution-feldspar interface. The intensity ratio measurements for aluminum and silicon at various pH values show no significant change for a variety of dissolution experiments. This invariance in the ratios is evidence to support the notion that no precipitation or leaching process is significant in the dissolution process of feldspar grains.

Previous discussion has emphasized the fact that XPS measurements can provide unique information regarding the oxidation state[6-9,16,17] and coordination geometry[20] for metal species adsorbed from aqueous solution. In this paper, the chemical nature of Pb(II) adsorbed on manganese dioxide and Co(II) on illite has been probed using XPS binding energy data and photoelectron peak characteristics. Special attention has been given to the satellite structure for Co(II).

EXPERIMENTAL

The materials used in this investigation were obtained from the following sources: illite, Fithian, Illinois; A.P.I. Standard #36, Illinois Geological Survey; and β-MnO_2, commercial supplier. The δ-MnO_2 sample was prepared as described previously.[9] A sample of pure PbO_2 was prepared as described by Counts et al.[4] A mixture of PbO_2 and PbO was prepared by heating lead foil at 200°C for one week in a sealed Pyrex tube containing approximately one atmosphere of O_2. The presence of PbO_2, and of PbO_2 with PbO using these preparative methods was confirmed by X-ray diffraction (XRD) methods. XRD data were obtained using nickel filtered CuKα radiation from a Picker X-ray diffractometer operated at 35 kV, 16 mamps and scanned at one degree per minute. All other materials were purchased from commercial sources.

The solutions were prepared using reagent grade metal salts dissolved in double distilled deionized water.[20] Constant ionic strength was maintained using $NaClO_4$ or NaCl and the pH of the solutions was adjusted to the desired value using 0.1 N NaOH or 0.1 N HNO_3. The adsorption experiments were conducted for 48 hours for Pb(II) adsorption on β- and δ-MnO_2. Cobalt (II) adsorption on illite was allowed to proceed for one week at constant controlled pH. The analysis of metal ion concentration was accomplished using atomic absorption techniques described earlier.[20,21]

XPS measurements were made with a Du Pont 650 spectrometer. Lead 4f spectra were determined using a Mg anode. Cobalt 2p spectra and binding energy determinations for Co adsorbed on illite were obtained using an Al anode because a portion of the oxygen Auger peak interfered with the measurements for all samples with low cobalt surface concentrations when using the Mg anode. Curve resolution procedures employed in this study were carried out with the GASCAP program.[22] Atom ratios were calculated from integrated photopeak intensities that had been corrected only by using the cross sections published by Scofield.[23] Binding energy results were calculated by measuring the C 1s binding energy for background carbon (BE = 284.6 eV). An alternative binding energy calibration procedure used an element of the substrate as the reference for the binding energy evaluation. The absolute binding energies for the substrate elements were determined using the 4f levels of gold or the 1s level of background carbon as a reference.[20-22] Samples were prepared for XPS analysis by spreading the powdered sample on double stick tape which covered a brass probe. Alternatively, samples were prepared by placing a few drops of acetone or ethanol suspension of the powdered sample on the brass probe and allowing the liquid to evaporate in air.

The XPS spectra did not change during the time required to obtain the data, using either the analog or multi-channel mode

for recording spectra. Spectra were obtained in such a manner
that the sample was exposed to the X-ray beam for less than three
hours. No evidence was obtained to suggest that X-ray induced
processes occurred during the measurements. The precision in the
binding energy results is ± 0.1 eV except for the manganese 3s
splitting which is ± 0.2 eV.

RESULTS AND DISCUSSION

Lead Adsorption on MnO_2

The principal theme of the present work is an attempt to use
XPS in identifying the oxidation state of metal ions adsorbed on
mineral surfaces, since previous studies[7,9] have provided evidence
for reduction/oxidation processes accompanying adsorption of
cobalt. To this end, the initial efforts were directed at determining the binding energies for authentic Pb(II) and Pb(IV)
compounds which would serve as guides in identifying the presence
of Pb(II) and Pb(IV). The binding energy results for authentic
lead oxides from this work and from other publications are summarized in Table 1. For lead oxides, it is apparent that the
literature values are not in agreement with respect to the absolute values for PbO and PbO_2 and with regard to the relative
differences between the Pb $4f_{7/2}$ binding energy values for PbO
and PbO_2. In three cases the binding energy for Pb(IV) is greater

Table 1.   Lead $4f_{7/2}$ Binding Energies for Lead-
           Containing Compounds

Binding Energy (eV)		ΔE Pb(IV)-Pb(II)	Ref.
PbO	PbO_2		
138.6	139.2	0.6	4
139.8	141.0	1.2	24
137.4	138.4	1.0	a
138.1	137.9	-0.2	25
137.6	137.2	-0.4	26
138.2	137.6	-0.6	27
138.0[b]		–	a

a This work

b Mixed PbO/PbO_2 oxide

than that for Pb(II), while for the other three measurements, the
Pb(IV) Pb $4f_{7/2}$ binding energy is only slightly lower than that
for Pb(II). For each group of binding energy measurements (i.e.
Pb(IV)>Pb(II) and Pb(II)>Pb(IV)), it appears that the absolute
values for the binding energies may differ as a result of calibration standards. On the other hand, the lower binding energy
for Pb in PbO_2 compared to the value for Pb in PbO has been
attributed to relaxation effects.[26] In the present work, the PbO
sample was a commercial powdered sample while the PbO_2 material
was prepared by controlled oxidation of lead foil.[4] In another
experiment, a mixed PbO_2/PbO lead surface was prepared as described
in the experimental section. The PbO, PbO_2 and the mixed PbO/PbO_2
samples were characterized by XRD measurements. An analysis of
the diffraction results suggests that the ratio of PbO/PbO_2 for
the mixed oxide sample is approximately 1:1. An analysis of the
XRD data for the PbO_2/Pb foil indicates that the maximum percentage
of PbO in the sample would be much less than 5%. The binding
energy results for the PbO and PbO_2 materials indicate that the
binding energy for Pb(IV) is greater than that for Pb(II). For
the mixed PbO/PbO_2 sample, the Pb $4f_{7/2}$ binding energy falls
between the value measured for the individual PbO and PbO_2
samples. In addition, the peak width at half maximum (PWHM) for
the mixed oxide sample was 1.8 eV while the corresponding values
for PbO and PbO_2 were 1.6 and 1.5 eV, respectively.

    In Figure 1 the curve resolved spectrum for the mixed
Pb(II)/Pb(IV) oxide sample is presented. The dotted curve is the
fit to the experimental curve (solid line). The Pb 4f components

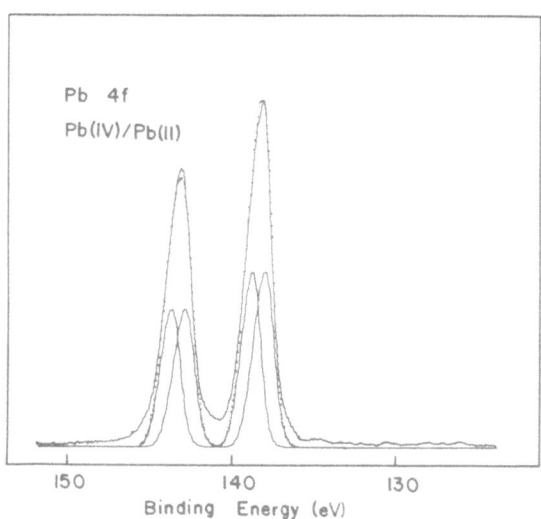

FIGURE 1   Curve resolved lead 4f XPS spectra for mixed lead oxide.

Table 2.   Binding Energy Results - Pb(II)/β- and δ-MnO$_2$

Sample	Concentr. Pb(mole)/m^2	Pb 4f$_{7/2}$ BE (eV)	PWHM (eV) 4f$_{7/2}$	Mn 2p$_{3/2}$ BE (eV)	Mn 3s Splitting (eV)	PWHM (eV) lower BE 3s peak
			β-MnO$_2$			
pure β-MnO$_2$	-	-		642.2	4.5	3.0
1	1.546x10^{-7}	137.8	1.7	642.2	4.9	3.6
2	2.725x10^{-7}	138.0	1.7	642.1	4.7	3.5
3	3.727x10^{-7}	138.0	1.8	642.2	4.8	3.6
4	4.409x10^{-7}	138.0	1.7	642.3	4.7	3.5
5	1.005x10^{-6}	138.0	1.8	642.3	4.9	3.8
6	1.359x10^{-6}	137.9	1.7	642.2	4.8	3.6
7	1.409x10^{-6}	137.9	1.8	642.5	4.7	3.7
8	1.995x10^{-6}	138.1	1.8	642.2	4.7	3.7
9	5.767x10^{-6}	137.7	1.7	642.2	4.9	3.6
10	5.992x10^{-6}	138.0	1.8	642.4	4.7	3.6
			δ-MnO$_2$			
pure δ-MnO$_2$	-	-		642.3	4.5	3.5
1	1.494x10^{-7}	137.9	1.8	642.2	5.1	
2	2.614x10^{-7}	138.0	1.8	642.4	5.0	3.7
3	2.091x10^{-6}	137.8	1.7	642.7	5.1	4.0
4	5.229x10^{-6}	138.0	1.7	642.4	5.0	4.1
5	1.046x10^{-5}	138.0	1.8	642.3	5.0	4.0
6	2.062x10^{-5}	138.0	1.8	642.4	5.1	4.0
7	5.197x10^{-5}	138.1	1.8	642.4	5.0	4.3
8	5.372x10^{-5}	137.9	1.7	642.4	5.1	4.3
9	1.519x10^{-4}	137.7	1.7	642.4	4.9	4.2
MnO				640.6*	5.8*	
γ-Mn$_2$O$_3$				641.7*	5.2*	
α-Mn$_2$O$_3$				641.9*	5.2*	
β-MnO$_2$				642.2*	4.6*	

*Data from Reference (28).

used to fit the experimental curve are also presented as solid
curves.   The binding energies and peak widths for the Pb(II) and
Pb(IV) components are equivalent to those for pure PbO and PbO$_2$,
respectively.   These results are in accord with the notion that
lead foil oxidized in a static oxygen atmosphere at 200°C produces

a mixture of PbO and PbO_2. Even though the XPS and XRD measurements differ in their sensitivity to the surface composition, the
results are in qualitative agreement in that approximately a 1:1
PbO/PbO_2 mixture is present on the foil oxidized at 200°C. For
the subsequent discussion the Pb $4f_{7/2}$ binding energy values for
Pb(IV) and Pb(II) are 138.4 and 137.4 eV, respectively.

    The binding energy values measured for lead (II) adsorbed on
β- and $δ-MnO_2$ are summarized in Table 2. The binding energies
were measured for adsorbed lead at a variety of surface concentrations to probe whether the extent of surface coverage might
influence the measured binding energy. Examination of the binding
energy results indicates that there is no variation, within
experimental error, of the Pb $4f_{7/2}$ binding energies for β- or
$δ-MnO_2$ samples as a function of lead concentration. (The measured
binding energy is also equivalent for lead present on $β-MnO_2$ and
on $δ-MnO_2$.) Thus it appears that the chemical nature of adsorbed
lead is identical on β- and $δ-MnO_2$ at all of the concentrations
investigated. The binding energy data are similar to the data
obtained for the mixed oxide foil in that the binding energies
cluster about the value 138.0 eV. It is also significant that
the measured PWHM is slightly greater than that for pure PbO_2 or
PbO. Such results indicate that lead may be adsorbed as a mixture
of Pb(II) and Pb(IV). To obtain an approximate evaluation of the
Pb(II)/Pb(IV) ratio, the Pb 4f photopeaks were curve resolved
using the GASCAP IV program. The input data included the binding
energy values for pure PbO (137.4 eV) and pure PbO_2 (138.4 eV)
and the PWHM for each of the 4f levels for each oxidation state.
To obtain a fit to the raw data, only the heights of the Pb(II)
and Pb(IV) 4f photopeaks were adjusted. A typical fit for the
lead 4f photopeaks is shown in Figure 2. From the curve resolved
Pb $4f_{7/2}$ spectra, it is estimated that the lead composition on
the MnO_2 surface is approximately 30% Pb(IV). The formation of
Pb(IV) on the MnO_2 surface is in accord with an earlier prediction[9]
that Pb(II) would be oxidized upon adsorption. From the present
results it is found that only partial oxidation occurs. Because
the model prediction[9] is based only on thermochemical considerations, it may be that kinetic factors influence the formation of
the higher oxidation state of lead. No effort has been made in
the present study to explore the kinetic aspects of Pb(IV) formation.

    In an effort to discover whether manganese might be the
oxidizing agent in the adsorption/oxidation process, the binding
energies for the Mn $2p_{3/2}$ and for the Mn 3s levels were measured.
The Mn 3s level produces two photopeaks due to multiplet splitting
(shown in Figure 3) and the splitting value is given in Table 2
along with the Mn $2p_{3/2}$ binding energies. The Mn $2p_{3/2}$ binding
energy values for the $Pb-MnO_2$ samples are equal to the values for

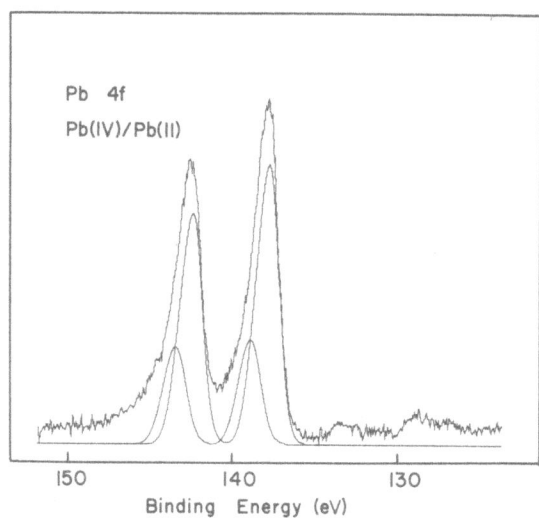

FIGURE 2   Curve resolved lead 4f XPS spectra for lead adsorbed on
β-MnO$_2$.

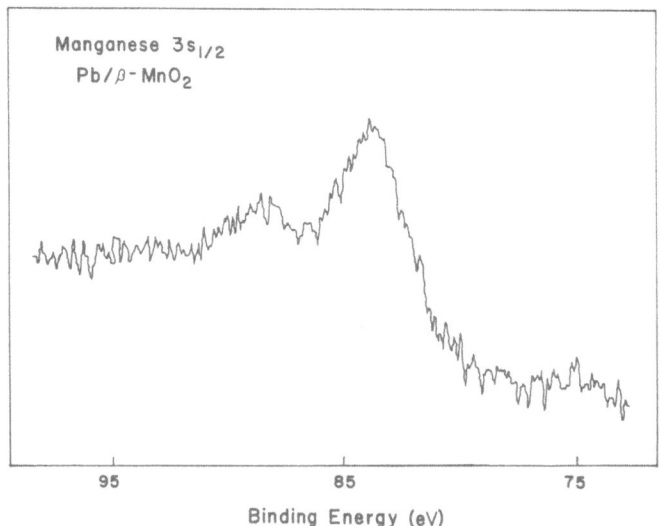

FIGURE 3   Manganese 3s XPS spectrum for lead adsorbed on β-MnO$_2$.

pure β- and δ-MnO$_2$. The measured Mn 2p$_{3/2}$ binding energy and the
Mn 3s splitting in this work are in good agreement with the
values as reported earlier.[28] The close agreement in the
Mn 2p$_{3/2}$ binding energies for the Pb-MnO$_2$ and the pure MnO$_2$
materials suggests that the oxidation state of surface manganese
is Mn(IV). However, a more convenient indicator of the manganese
oxidation state is the extent of Mn 3s splitting. From the data

in Table 2, it is noted that the Mn 3s splitting increases as the oxidation state of Mn changes from Mn(IV) to Mn(II) for authentic manganese oxides. For the Pb/MnO$_2$ samples, the Mn 3s splitting for both β- and δ-MnO$_2$ samples is greater than that for the pure oxides. These Mn 3s splitting values fall between the 4.5 eV value for β- and δ-MnO$_2$ and the 5.2 eV value for γ- and α-Mn$_2$O$_3$. In addition, the PWHM values for the lower binding energy peak of the 3s multiplet split double is about 0.5 eV wider than that for either β- or δ-MnO$_2$ (see Table 2). These results indicate that perhaps some reduction of Mn(IV) has occurred during the adsorption process. In the study of Co(II) adsorption on MnO$_2$,[9] the XPS results indicated that all adsorbed cobalt was present as Co(III) and it was postulated that dissolved oxygen was the oxidizing agent. No evidence for the presence of surface Mn(III) was obtained.[9] In the present study, it is not possible to rule out the role of oxygen as an oxidizing agent during the adsorption experiments.

Chemical reactions consistent with either oxidation/reduction process, are summarized in Equations (1) and (2), where PbO$_{2(s)}$ is written to represent Pb(IV).

$$Pb^{2+}_{(aq)} + 1/2O_{2(g)} + H_2O_{(\ell)} \rightleftharpoons PbO_{2(s)} + 2H^+_{(aq)} \tag{1}$$

$$Pb^{2+}_{(aq)} + 2MnO_{2(s)} + H_2O_{(\ell)} \rightleftharpoons PbO_{2(s)} + Mn_2O_{3(s)} + 2H^+_{(aq)} \tag{2}$$

It is possible that the mode of oxidation/reduction is different for cobalt (II) and lead (II) adsorption on MnO$_2$ surfaces. Such a difference might explain why adsorbed oxidized lead may be limited by the surface concentration of Mn(IV) and by the availability of surface sites for oxidation. Our research efforts at this point have not addressed this question, although experiments are in progress in an attempt to discover if there is a difference in the mechanism of oxidation for these divalent ions on MnO$_2$ surfaces.

The quantitative analysis of metal ions adsorbed on various substrates using XPS has been demonstrated in the past.[15,18,29-31] Although the principal thrust of this study was not quantitative in nature, it was found that a linear relationship existed between the surface concentration of lead (mol·Pb/m^2 surface area) and the integrated intensity ratio of Pb to Mn. Peak areas for the Pb 4f$_{7/2}$ and for the Mn 2p$_{3/2}$ photopeaks corrected for the photoelectric cross section were used to calculate the ratio. No effort was made to account for differences in electron escape depth. These results are shown graphically in Figures 4 and 5

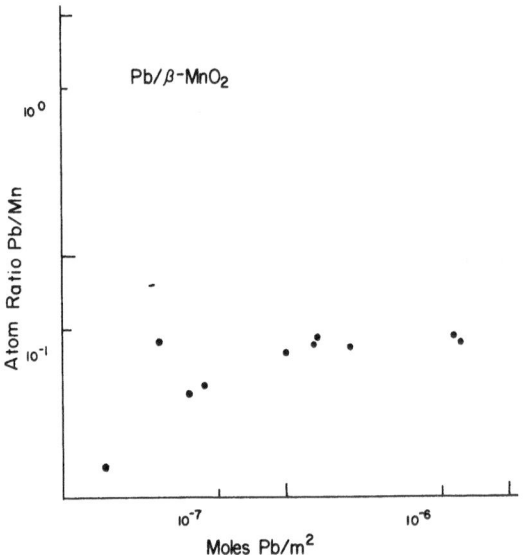

FIGURE 4   Surface lead concentration vs. Pb/Mn XPS ratio for
           β-MnO$_2$.

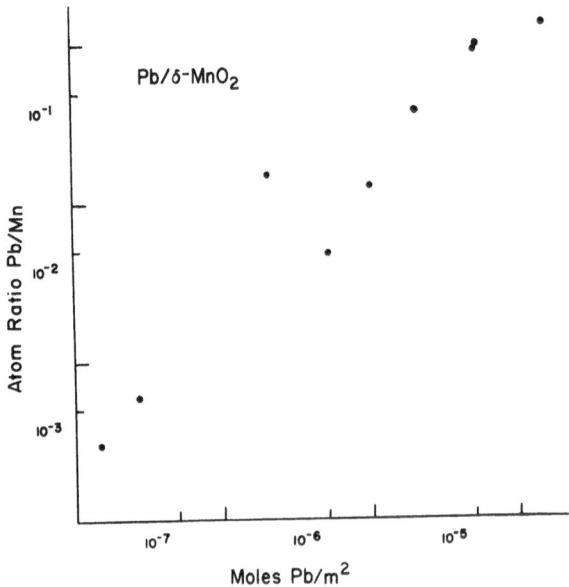

FIGURE 5   Surface lead concentration vs. Pb/Mn XPS ratio for
           δ-MnO$_2$.

for β-MnO$_2$ and for δ-MnO$_2$, respectively. A linear correlation
between the lead surface concentration and the Pb/Mn ratio is
noted for δ-MnO$_2$, while for β-MnO$_2$ a limiting value for the Pb/Mn
XPS ratio is observed at approximately 1.0 x 10^{-6} mol·Pb/m^2.

It is also noteworthy that the Pb/Mn XPS ratio is not equal for
equal surface concentrations for Pb on the β- and δ-MnO_2 samples.
This point is emphasized by considering, for example, that the
Pb/Mn XPS ratio is about 30 times greater for β-MnO_2 than for
δ-MnO_2 when considering data for each #1 sample of MnO_2.   At
present no explanation can be offered for this dramatic difference
although differences in surface morphology and pore structure
could be responsible for this phenomena.

Cobalt (II) Adsorption on Illite

The adsorption of cobalt on clay minerals has been studied
by a number of workers.[5-7,9,13,14]   Adsorption of Co(II) on metal
oxides yields surface cobalt hydroxide[5,6] while adsorption of
aqueous Co(II) on chlorite at pH 3 and 7 does not form the
hydroxide.[7] Adsorption of Co(III) amine complexes in aqueous
solution leads to hydrolysis of the complex with an aqua cobalt (II)
species adsorbed on the surface. The investigation reported here
is an XPS examination of the nature of adsorbed Co(II) studied as
a function of solution pH.

The Co 2p spectra for adsorption at various pH values are
shown in Figure 6. The measured binding energies for the $2p_{3/2}$
level as well as the Co $2p_{1/2}$ - Co $2p_{3/2}$ splitting energies are
given in Table 3. The XPS spectrum measured for cobalt adsorbed
at pH 5 reveals only a low intensity peak for the Co $2p_{1/2}$ level
and a broad signal in the Co $2p_{3/2}$ region. The broadness of the
peak is attributed to the presence of an iron Auger peak and thus

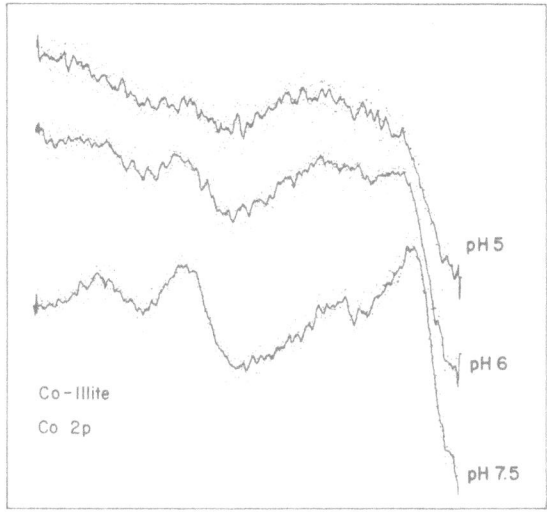

FIGURE 6   Cobalt 2p XPS spectra for Co(II) adsorbed on illite.

Table 3. Co $2p_{3/2}$ Binding Energies for Co(II) Adsorbed on
          Illite at Various pH Values

pH	Binding Energy (BE) (eV)	ΔBE (Co $2p_{1/2}$-Co $2p_{3/2}$)
5.0	–	–
6.0	782.0 ± 0.2	15.8
7.0	781.6 ± 0.1	15.9
7.5	781.5 ± 0.1	16.0
7.8	781.1 ± 0.1	15.9
8.0	781.1 ± 0.1	15.9
10.0	781.1 ± 0.1	15.9
Co(OH)$_2$	781.0	15.9*
CoO	780.0	16.0*
Co(II)/ chlorite	782.1	16.0*

*Data from References (6) and (7).

no binding energy results were obtained for the pH 5 sample. At
pH 6, the $2p_{1/2}$ photopeak and its associated shake-up satellite
are visible and the main photopeak for Co $2p_{3/2}$ is just discern-
ible. At pH 7 and at other pH values reported in this study,
both cobalt 2p peaks were easily recorded.

The presence of shake-up satellite features and the Co $2p_{1/2}$
- Co $2p_{3/2}$ splitting of approximately 16 eV (see Table 3) are
supporting evidence for the presence of Co(II) on the illite
surface. At pH 6, the Co $2p_{3/2}$ binding energy is equal to that
measured for Co(II) adsorbed on chlorite[7] and on kaolinite and
ripidolite,[32] while the values at pH 7.8, 8.0, and 10.0 are
equivalent to that for Co(OH)$_2$. At pH 7.0 and 7.5, a Co $2p_{3/2}$
binding energy between that for cobalt at pH 6 and that for
Co(OH)$_2$ is recorded. The interpretation of these results indicates that the chemical nature of adsorbed Co(II) is dependent on
pH and that if adsorbtion is carried out at pH values in the
region where Co(OH)$_2$ can form, this is the favored reaction. At
the intermediate pH values an important cobalt (II) species is
Co(OH)$^+$.[33] The binding energy values for pH 7 and 7.5 could
arise for a combination of cobalt species including aqua Co(II),
Co(OH)$^+$, and Co(OH)$_2$. Although curve resolution analyses of the
Co 2p spectra could provide results that might be useful in
elucidating the chemical nature of adsorbed cobalt (II), it is
impractical in the present case due to the presence of satellite
features and the potential presence of the iron Auger peak. The

most reasonable interpretation of the present results is that an aqua Co(II) species is adsorbed at pH 6, $Co(OH)_2$ is formed at pH values 7.8 - 10.0, and any combination of $Co(OH)_2$, Co(II)(aq) and $Co(OH)^+$ may be present at pH values in the 6 - 7.8 range.

ACKNOWLEDGEMENTS

We thank the National Science Foundation and the Office of Water Resources Technology for support of this research. Thanks are also extended to Professor J. Craig of the VPI Department of Geology for his help with the X-ray diffraction measurements and interpretation.

REFERENCES

1. S.L. Swartzen-Allen and E. Matijevic, Chem. Rev. 74:385 (1974).
2. M.M. Mortland and F.C. Farmer, ed., "International Clay Conference 1978", Elsevier, Amsterdam (1978).
3. G.P. Glasby, ed., "Marine Manganese Deposits", Elsevier, Amsterdam (1977).
4. M.E. Counts, J.S.C. Jen, and J.P. Wightman, J. Phys. Chem. 77:24 (1973).
5. P.H. Tewari and W. Lee, J. Colloid Interface Sci. 55:77 (1975).
6a. P.H. Tewari and N.S. McIntyre, AIChE Symposium Series 71:150 (1975).
6b. N.S. McIntyre and P.H. Tewari, J. Colloid Interface Sci. 59:195 (1977).
7. M.H. Koppleman and J.G. Dillard, J. Colloid Interface Sci. 66:345 (1978).
8. J.W. Stucki, C.B. Roth and W.E. Baitinger, Clays Clay Min. 24:289 (1976).
9. J.W. Murray and J.G. Dillard, Geochim. Cosmochim. Acta 43:781 (1979).
10. R.A. Schoonheydt, F. Velghe and J.B. Uytterhoeven, Inorg. Chem. 18:1842 (1979).
11. F. Velghe, R. Schoonheydt and J.B. Uytterhoeven, Clays Clay Min. 25:375 (1977).
12. J. Chaussidon, R. Clavet, J. Helsen and J.J. Fripiat, Nature 196:201 (1962).
13. J.J. Fripiat and J. Helsen, Clays Clay Min. 14:163 (1966).
14. M.H. Koppelman and J.G. Dillard, "International Clay Conference 1978", M.M. Mortland and F.C. Farmer, eds., Elsevier, Amsterdam (1978). p. 153.
15. G.M. Bancroft, J.R. Brown and W.S. Fyfe, Chem. Geol. 25:227 (1979).

16. J.H. Lunsford, P.J. Hutta, M.J. Lin and K.A. Windhorst, Inorg. Chem. 17:606 (1978).

17. J.W. Stucki and C.B. Roth, Soil Sci. Soc. Amer. J. 41:808 (1977).

18. G.M. Bancroft, J.R. Brown and W.S. Fyfe, Chem. Geol. 19:131 (1977).

19. R. Petrovic, R.A. Berner and M.B. Goldhaber, Geochim. Cosmochim. Acta 40:537 (1976).

20. M.H. Koppelman and J.G. Dillard, Clays Clay Min. 25:457 (1977).

21. M.H. Koppelman and J.G. Dillard, ACS Symp. Ser. 18:186 (1975).

22. J.H. Burness, J.G. Dillard and L.T. Taylor, J. Amer. Chem. Soc. 97:6080 (1975).

23. J.H. Scofield, J. Electron Spectrosc. Relat. Phenom. 8:129 (1976).

24. W.E. Swartz and D.M. Hercules, Anal. Chem. 43:1774 (1971).

25. W.E. Morgan and J.R. Van Wazer, J. Phys. Chem. 77:964 (1973).

26. K.S. Kim, T.J. O'Leary and N. Winograd, Anal. Chem. 45:2214 (1973).

27. J.M. Thomas and M.J. Tricker, J. Chem. Soc. Farad. Trans. II 71:329 (1975).

28. M. Oku, K. Hirokawa and S. Ikeda, J. Electron Spectrosc. Relat. Phenom. 7:465 (1975).

29. D.M. Hercules, L.E. Cox, S. Onisick, G.D. Nichols and J.C. Carver, Anal. Chem. 45:1973 (1973).

30. M. Czuha and W.M. Riggs, Anal. Chem. 47:1836 (1975).

31. G.M. Bancroft, J.R. Brown and W.S. Fyfe, Anal. Chem. 49:1044 (1977).

32. J.G. Dillard, M.H. Koppelman and A.B. Emerson, unpublished data, 1978-79, VPI & SU.

33. C.F. Baes and R.E. Mesmer, "The Hydrolysis of Cations", Wiley-Interscience, New York (1976). pp. 238-242.